THE WORLD FUTURES GENERAL EVOLUTION STUDIES
A series edited by Ervin Laszlo
The General Evolution Research Group
The Academy for Evolutionary Management
and Advanced Studies, Fulda, Germany

*VOLUME 1*
**NATURE AND HISTORY: THE EVOLUTIONARY APPROACH FOR SOCIAL SCIENTISTS**
Ignazio Masulli

*VOLUME 2—KEYNOTE VOLUME*
**THE NEW EVOLUTIONARY PARADIGM**
Edited by Ervin Laszlo

*VOLUME 3*
**THE AGE OF BIFURCATION: UNDERSTANDING THE CHANGING WORLD**
Ervin Laszlo

*VOLUME 4*
**COOPERATION: BEYOND THE AGE OF COMPETITION**
Edited by Allan Combs

*VOLUME 5*
**THE EVOLUTION OF COGNITIVE MAPS: NEW PARADIGMS FOR THE TWENTY-FIRST CENTURY**
Edited by Ervin Laszlo and Ignazio Masulli
with Robert Artigiani and Vilmos Csányi

*VOLUME 6*
**THE EVOLVING MIND**
Ben Goertzel

*VOLUME 7*
**CHAOS AND THE EVOLVING ECOLOGICAL UNIVERSE**
Sally J. Goerner

*VOLUME 8*
**CONSTRAINTS AND POSSIBILITIES: THE EVOLUTION OF KNOWLEDGE AND THE KNOWLEDGE OF EVOLUTION**
Mauro Ceruti

*VOLUME 9*
**EVOLUTIONARY CHANGE: TOWARD A SYSTEMIC THEORY OF DEVELOPMENT AND MALDEVELOPMENT**
Aron Katsenelinboigen

This book is part of a series. The publisher will accept continuation orders which may be cancelled at any time and which provide for automatic billing and shipping of each title in the series upon publication. Please write for details.

# Evolutionary Change

# Evolutionary Change

## Toward a Systemic Theory of Development and Maldevelopment

**Aron Katsenelinboigen**
*University of Pennsylvania*
*Philadelphia*

**Gordon and Breach Publishers**
Australia • Canada • China • France • Germany • India • Japan • Luxembourg • Malaysia •
The Netherlands • Russia • Singapore • Switzerland • Thailand • United Kingdom

Copyright © 1997 OPA (Overseas Publishers Association) Amsterdam B.V. Published in The Netherlands under license by Gordon and Breach Science Publishers.

All rights reserved.

No part of this book may be reproduced or utilized in any form or by any means, electronic or mechanical, including photocopying and recording, or by any information storage or retrieval system, without permission in writing from the publisher. Printed in India.

Amsteldijk 166
1st Floor
1079 LH Amsterdam
The Netherlands

---

British Library Cataloguing in Publication Data

Katsenelinboigen, Aron, 1927-
   Evolutionary change : toward a systemic theory of development and maldevelopment. - (World futures general evolution studies ; v. 9)
   1.Sociobiology
   I.Title
   304.5

ISBN 90-5699-529-4

*To*

*My grandmother* Rebecca

*My mother* Ida

*My relatives*

    Bluma Balter-Itsikovich, Elisabeth Feldman,
Sarra Krimotat, Gregory Stolerman and Efim Yusim

*My friends*

    Ida Abramovskaia-Shats, Alexandra Ackoff-Makarova,
Alexander Ash, Ilia Blokh, Emmanuel Braverman,
Michail Breev, Vladimir Dilman, Roland Dobrushin,
Ivan Elagin, Frances Finkelstein, Elizabeth Flower,
Emma Gurevich-Feld, Nadia Krasnoselsky,
Roberta Hall-Luxenberg, Leonid Kantorovich,
Tamara Libinzon-Peschansky, Wolfgang Pflugfelder,
Irene Rubin, Itzhak Sankowsky, Boris Shragin,
David Solomons, Michail Sonin, Leonard Starobin, and
Natalia and George Vladutz

*victims of cancer*

# Contents

| | | |
|---|---|---|
| *Introduction to the Series* | | ix |
| *Acknowledgments* | | xi |
| Introduction | | 1 |

**ONE**     **An Analogy Between Socio-Economic and Biological Mechanisms of Change**     **19**

     **1** Economic Mechanisms of Change and Biological Evolution     21

     **2** Mechanisms of Change in Socio-Political Systems and Biological Evolution     39

**TWO**     **Evolutionary Mechanisms of Change: Normal Case**     **49**

     **3** The Mechanisms of Biological Change—General Characteristics     51

     **4** Two Classes of Evolutionary Mechanisms of Change     75

     **5** Internal Mechanisms of Change     85

     **6** Special Features of the Somatic Mechanism of Change     109

     **7** Some Features of the Germatic Mechanism of Change: The Origins of Sex Differentiation     121

**THREE**     **Evolutionary Mechanisms of Change: Pathology**     **145**

     **8** First Steps En Route to a New Concept of Cancer     147

     **9** Characteristics of Cancer     171

Conclusion     191

References     199

Index     213

# Introduction to the Series

*The World Futures General Evolution Studies* series is associated with the journal *World Futures: The Journal of General Evolution.* It provides a venue for monographs and multiauthored book-length works that fall within the scope of the journal. The common focus is the emerging field of general evolutionary theory. Such works, either empirical or practical, deal with the evolutionary perspective innate in the change from the contemporary world to its foreseeable future.

The examination of contemporary world issues benefits from the systematic exploration of the evolutionary perspective. This happens especially when empirical and practical approaches are combined in the effort.

*The World Futures General Evolution Studies* series and journal are the only internationally published forums dedicated to the general evolution paradigms. The series is also the first to publish book-length treatments in this area.

The editor hopes that the readership will expand across disciplines where scholars from new fields will contribute books that propose general evolution theory in novel contexts.

# Acknowledgments

I wish to express my gratitude to:

Mark Adams, Michael Blagosklonny, Edward Budovsky, Valery Chalidze, Richard Cooper, Zinovy Chereisky, Vincent Cristofalo, Giulio D'Angio, Vladimir Dilman, Galina Filolenko, Michael Halpern, Norman Gross, Roland Kallen, Alfred Knudson, Vladimir Kochmashev, Alexey Kondrashev, Hilary Koprowski, Vadim Levin, Irene Lubensky, Nadia Lumelsky, Peter Nowell, Vilen Peschansky, Anatol Rapoport, Lev Rozonoer, Stanley Salthe, James Shapiro, William Telfer, Yuri Vasiliev, Michael Vilenchik and Vera Zubarev for valuable discussions.

It goes without saying that they bear no responsibility for the views expressed here. The only thing my colleagues may be held accountable for is disregarding mankind and spending time talking to me rather than doing their own valuable research.

I am especially grateful to Professor Anatol Levin and Igor Israeltian for sharing their insights into the problems of cancer, and Richard von Sternberg for valuable discussions and his thorough reading of the manuscript.

Helena Dubrovina and Janne Hunter, librarians at the Wistar Institute, and Sherry Morgan, librarian of the Biomedical Library of the University of Pennsylvania, have been very helpful in compiling the bibliography of works dealing with the evolutionary mechanisms of change and cancer.

Many thanks to the University of Pennsylvania Research Fund and its former head Barry Cooperman for his concern and financial support.

I appreciate the efforts of my son Alexander in translating this work into English, and of Marge Weiler and Idith Lapowsky for their editing.

# INTRODUCTION

## 1. MAIN HYPOTHESIS

Truly amazing advances in biology, particularly molecular biology, have opened new engineering possibilities to shape the future of mankind, both in terms of "improving" the human species as well as eradicating all kinds of pathological manifestations of biological development. However, all of these promising interventions are fraught with grave dangers. The pitfalls stem primarily from the local nature of changes that are introduced, meaning that the impact of the intervention upon the overall development of the biological system is not fully taken into account. As we struggle to improve or rectify the well-being of individual human beings, we are apt to harm the population as a whole and impair the development of the human species. In addressing all of these issues, it is essential to recognize the integral characteristics of diverse biological mechanisms of change brought about by nature.

Biologists are only now starting to pay attention to this class of the mechanisms of change. The great majority of the profession persists in reducing evolution to a uniform mechanism of random mutations. At the same time, certain order exhibited by the evolutionary process of change presumes, and plausibly so, certain variety in the actual mechanisms of change. Of course, this scheme does not ignore the role of the random mechanism of change.

My aim in this book is to explore new avenues of research into individual biological mechanisms of change as well as biological mechanisms of change in their totality, i.e., as they fit in the general dynamics of biological development. The book makes extensive use of the design approach.[1]* Discovery is really a result of testing the set of constructed models against one model, which, based on all the

---
*Square brackets indicate references at the end of the chapter

experimental data, is taken to be the model of the external, so called real world. The principle of design does not deny the role of the real world in the creative being as far as stimulating creative ideas, testing their validity in this real world, and their actual implementation (also using the example of certain pathological manifestations of these mechanisms of change).

My principal conjecture in the present work is the following: there is a somatic mechanism of change operating in multicellular organisms. It should be stressed that in speaking of somatic change I mean the entire mechanism of somatic change, i.e., the interaction of groups of specialized cells that undergo many stages of development. Change (mutations) within individual somatic cells confined to the host tissue is another matter.[1]

I set the mechanism of somatic change apart because its normal function and operation are rather poorly explored. While my focus is on the somatic mechanism of change, I would certainly not rule out its connection with the mechanism of change based on specialized reproductive cells - germ cells. In fact, I believe there is a certain relationship between these two mechanisms. Taking into account that "the genome of the somatic cells is the same as the genome of the germ cells"[2] (p.255), considerations expressed in the present work regarding the mechanism of change via somatic cells are largely pertinent to germ cells.

The following phenomena seem to corroborate my working assumption regarding the role of the somatic mechanism of change:

in terms of the evolutionary time table, the somatic mechanism of reproduction preceded reproduction based on germ cells;

the emergence of new reproductive methods on the biological scene did not eradicate previous modes of reproduction, but merely limited their scope of operation: reproduction based on somatic cells is still important to the development of living creatures. It may be argued that this mechanism is archaic, that it is not autonomous but plays a subordinate role, and that the changes it induces are not passed on to the progeny.

In pursuing my main theme, I explore two related issues having to do with somatic and germ cells as well as their interactions.

The first problem is the sources of change, meaning the interplay between internal and external sources of change and how ordered these changes are.

The prevailing opinion among biologists is that the sources of change are primarily external - radiation, chemicals, etc. Moreover, these changes are of random nature and are manifest in all kinds of damage (mistakes, disorders) in the genome. The utility of the damage in terms of the evolutionary process is to be determined by his majesty the process of selection. I would not discard these methods of the creation of new structures and the process of selection as the ultimate judge of the evolutionary worth of the innovations. Nonetheless, another very plausible hypothesis is that the genome supports more ordered processes of change, and it is at this earlier stage that the potential benefits, or damages, induced by the change can be ascertained. In view of the above, the term innovator shall only refer to those beings that exhibit a trend toward development prior to them being affected by the pressures of natural selection.

A number of biologists subscribe to the notion of internal mechanisms of change. The definition of this category is given by Lancelot White[3] in a following way:

> "INTERNAL (OR DEVELOPMENTAL) SELECTION
> The internal selection of mutants at the molecular, chromosomal, and cellular levels, in accordance with their compatibility with the internal coordination of the organism. The restriction of the hypothetically possible directions of evolutionary change by internal organizational factors." (p.vii)

The existence of internal mechanisms of change has been the subject of an extensive and long-running controversy. Without going into the history of science, perhaps the views descend from Bergson's[4,5] ideas of creative evolution, his *elan vita,* and the so-called *vitalist* school[6] are the principal forces in championing the existence of internal mechanisms of change. While the vitalist theory of an inner life-force irreducible to chemical or physical forces is highly controversial, its great achievement lies in its belief in the existence of an internal mechanism which makes life autonomous and self-evolving. Just as Mendel's statement of the existence of genes – long before genes themselves were discovered, had a profound influence on biology – the vitalist notion of an internal mechanism of self-evolving life can be quite conducive to current research armed with modern tools for uncovering the mechanisms of change.

The history of ideas regarding the internal mechanisms of change and described in White's book reveals that biologists who have recognized its existence tend to focus on the non-trivial links between the internal and the external (Darwinian) mechanisms of change. The internal mechanism of change itself was regarded as a black box, meaning its analysis was limited to phenomenological observations. The probe into the actual structure of the internal mechanism has only started in the recent past as biologists began to recognize the dynamic structure of the genome. We have embarked upon a long and fascinating journey into the internal mechanisms of change manifested in numerous discoveries, among which are *selfish genes* possessing definitive *linguistic patterns*, *transposons* (firstly discovered by Barbara McClintock, but deemed unimportant for a long time), a *computer based on the DNA molecule* designed by Leonard Adleman, etc.

In terms of a process, external sources of change operate *from the end*, meaning that they reflect external conditions (environment). Internal sources of change operate primarily *from the beginning*, creating predisposition toward future development. The joint action of these two sources of change creates the so called *tunnel process* in biological evolution. Possibly the genetic changes induced at the beginning are accumulated in the genome prior to their ultimate expressions in the phenotype.[7] Consequently, paleontologists need not seek all the intermediate changes that might be redundant or even harmful in the evolution of a phenotype of a new species. It is further surmised that internal mechanisms of change may resemble metaprograms, or programs for changing other programs, which, in turn, govern the development of an organism. Note the distinction between *regulatory* and *repair* structures incorporated into the program, as well as a program for changing the program. A program in the genome may carry regulatory, similar to the steam valve in Watt's steam engine, as well as repair genes that manage the genome's operation and preserve the program's integrity. These segments of the program should not be confused with programs for changing other programs, i.e., programs that orchestrate the performance of adjacent level programs. A formalized analysis of a similar kind of internal mechanism of change can be found in Robert Rosen's book (part 10 C, pp. 248-252).[8]

Chapter 5 explores in greater detail the modern-day discoveries in molecular biology that make the existence of an internal mechanism of change quite plausible.

Now a few words about the second issue related to the principal topic – the emergence of germ cells and their interaction with somatic cells.

It seems that change can be more effectively implemented via the germ cells if the genetic material which defines the development of a new organism is collected in one compact space. Over the course of evolution, germ cell based reproduction has undergone fundamental shifts. Evolution first gave rise to sexual cells, and then separated the carriers of these cells.

There is no agreement among biologists as to the origins of the sexes. One necessary condition, frequently deemed sufficient, for the formation of the sexes is their ability to create diversity of genetic combinations generated by crossing over between homologous chromosomes. I have tried to elaborate a general approach to sexual differentiation, not necessarily limited to two sexes, and to show that specific functional attributes of compatible organisms, directly participating in crossing, represent another condition needed to define different crossing types. It cannot be ruled out that change is not merely limited to germ cells, but also takes place in somatic cells. The latter may eventually migrate to germ cells via the female reproductive system, which is easily penetrated by somatic cells (at least pathological ones). On the other hand, invasion of the male reproductive organs by somatic cells is problematic.

By focusing on the less explored mechanisms of biological dynamics, the theory proposed here might help future research into the mechanisms of evolution, both under normal and pathological conditions. I hope to further the understanding of the somatic mechanism of change in the norm by focusing on its pathological manifestations. My basic hypothesis is that cancer represents a pathological attempt to reconstruct an organism via the mechanism of somatic change. The omnipresence of cancer - from plants to various cells in complex creatures - seems to corroborate the one condition necessary to make the above claim, namely that a pathological systems-oriented dysfunction in the mechanism of change is universal to all life.

Furthermore, cancer cells might turn out to be innovator cells of the radical variety which flourish under a weakened immune system. Although there are obvious external sources of cancer, I believe the main causes of this pathology are rooted in, or at least linked to, the internal mechanism of change. The similarity between transposons, which I beleive belong to the latter mechanism, and cancer causing virusis can be viewed as a minor confirmation of the above statement. During the last two or three decades, we have witnessed unprecedented progress in molecular biology which has shed light on the various stages of the malignant process: the role of telomeres and telomerase in immortalizing the cancer cell, malfunctioning genes responsible for slowing down cell growth, patterns to metastasis, etc. Recent research has also provided an opportunity to probe the normal process of somatic change. However general theories of cancer are lacking.[2] Biologists involved in cancer research tend to focus on isolated features of the disease, slighting more global concepts of this category of disorders. The empirical branch of cancer research is fueled by one very important consideration, namely that even if one single stage of cancer can be fully dissected and controlled, then the progress of the disease can be stopped. In other words, in order to destroy something, it is sufficient to eliminate one necessary condition of its existence.

It seems that the lack of comprehensive understanding of the nature of cancer will make it very difficult to eliminate even one of its necessary conditions. I shall elaborate upon this statement in the main body of the book. To reiterate, what distinguishes the concept of cancer expounded in this book is its totality.

I believe the proposed approach that attempts to explain cancer in its totality will benefit if we view cancer in terms of the general development of life, i.e., immerse cancer in the general evolutionary framework. Unfortunately, the evolutionary approach to cancer has not been adequately explored. In fact, theories of cancer rooted in the evolutionary framework tend to focus upon the *justification* of cancer and its devastating effects, thus fostering a conflict between the positive role of cancer in terms of evolution and the need to eradicate it at the microlevel (single organism).[3] My approach is predicated upon the notion of cancer as a pathology that ought to be overcome, both at the micro and the evolutionary levels. Recent progress in fighting cancer is fraught with certain dangers arising mainly from a lack of elaborate theories of cancer in terms of general evolution, thus pitting progress at the level of an organism against that of the species as a whole.

Thus, I believe the above argument is grounds for the following hypothesis: cancer belongs to a class of systemic diseases stemming from the malfunction of the normal mechanisms of change in the cell. The great majority of disorders, including infectious diseases, poor organ performance, etc., stem from the dysfunction of a normal mechanism that merely reproduces a cell.

To summarize, the present book contains four classes of problems: the axis problem is the somatic mechanism of change with three other problems rotating around it - internal mechanisms of change, sexual reproduction, and cancer.

In terms of content, the present work resembles a monograph.[4] It is written in the key of contemporary ideas from systems theory, adorned with ideas expressed in the old-fashioned style of natural philosophy.

In keeping with the systems approach, the study of biological mechanisms of change ought to be immersed in the more general system which, in this case, is the universe.

A considerable number of works[9, 10] have surfaced that attempt to examine biological evolution from a more general perspective, namely in terms of its link with the evolution of the inorganic word as well as its place in the social evolution of mankind.

Scholars of the holistic evolution of the universe strongly support the notion of a certain order governing the evolutionary process and reject the idea of randomness as the major driving force.

As noted by Ervin Laszlo:

> "It is reasonable to conclude that pure chance did not (and does not) dominate the evolutionary process: there must also be a significant degree of non-randomness.
> ...Given the ordered complexity that meets our eye, the reasonable assumption is that, somehow, preferential interconnections must exist in nature."[11] (pp.4-5)

The described general concept of cancer is made more concrete through the use of a new discipline termed General theory of evolution that emerged at the beginning of 1990s.

The champion of the new discipline, and one actively involved in promoting publications in the field, including editing World Futures: The Journal of General Evolution, was Ervin Laszlo[12], well known for his work in general systems theory. A major contribution to the development of this field was made by Ilya Prigogine.[13] One of the cornerstones of the new discipline is Prigogine's concept of nonequilibrium thermodynamics. Its applications extend beyond the realm of physics and chemistry to the evolutionary processes in biology, social systems, etc. Other interesting ideas of interdisciplinary significance contributing to the development of the general theory of evolution were expressed by Vilmos Csáni.[14]

The guiding principle of the general theory of evolution is the exploration of the general features of complex dynamic systems that exhibit development and the emergence of new structures. I believe that many of the ideas elaborated on in the present book that are pertinent hinge upon this guiding principle, e.g. the relationship among survival, viability, growth, and development, the interplay between the emergence of new entities, (expansion of diversity) and order, the tunnel process, evolutionary goals - to list a few.

While much of the research on the general theory of evolution examines the general characteristics of systems' evolution, a number of authors have probed specific systems, such as biological, social and economic ones.[15] Therefore, my analogies between the mechanisms of change in biological and socio-economic systems, as well my general deliberations on the nature of change in biological systems, also conform to the spirit of the general theory of evolution. In fact, the following topics touch upon new facets:

a) deliberations on the biological mechanisms of change (specific ones, as well as their collective behavior) that were initially based on somatic cells, then on a specialized reproductive cell, which, in turn, was transformed into differentiated germ cells with a subsequent separation of host/carriers of these cells;

b) my hypothesis that such a systemic disease as cancer is a manifestation of a pathological operation of the mechanism of change which appeared over the course of evolution; furthermore, based on this pathology, one can try to reconstruct certain less explored normal mechanisms of biological change that touch upon new facets of the evolutionary process and may contribute to the development of new branches of the general theory of evolution.

## 2. MECHANISMS OF EVOLUTIONARY CHANGE: IS THERE ROOM FOR NEW IDEAS?

Having done some reading on the subject of evolution, I have come to the conclusion that the basic research into the mechanisms of biological change addresses the following issues:

1) Is there one uniform mechanism of change, or a variety of such mechanisms?

2) Does change evolve in small disjoint steps (piece-meal), i.e., in the evolutionary manner, or does it unfold in broad comprehensive stages, i.e., in the revolutionary manner?

3) Are changes that take place random, or do they exhibit some sort of order?

4) Assuming there is some order in genetic changes, are these changes governed by rules pertaining to individual parts of an organism, or to holistic development?

5) Is there a special internal mechanism of change within the genetic structure, or is the genetic structure geared entirely to reproduction of the same organism? If so, are changes due to external perturbations?

6) Does as much change take place in somatic cells as takes place in germ cells?

# INTRODUCTION

7) What are the reasons for one germ cell (*spore*) that is sufficient to produce an organism, to evolve into gametes - specialized reproductive cells that must fuse to produce an offspring - and the subsequent separation of the carriers of these cells?

8) Are there players who intervene forcibly in the process of change, or does change evolve based on local interactions of equal participants?

The following dichotomies represent a concise summary of research into the aforementioned problems of biological mechanism of change:

A. 1 - uniformity, 2- diversity;
B. 1 - small disjoint steps, 2 - big broad steps;
C. 1 - randomness, 2 - order;
D. 1 - only via a genome, 2 - organism as a whole;
E. 1 - special internal mechanism of change, 2 - disturbances in genome structure;
F. 1- somatic cells, 2 - germ cells;
G. 1 - recombination, 2 - different functions;
H. 1 - vertical, 2 - horizontal mechanisms.

While the above parameters are not completely independent, each one underscores a specific aspect of development.

Figure 1 is based on these dichotomous variables. The cells, generated in a deductive manner, correspond with the various theories of evolutionary change. There are 256 different theories possible under the outlined scheme.

Generally speaking, the prevailing theories which conform to the scientific mind-set answer the eight questions posed above in the following way: (1) the mechanism of change is uniform, (2) changes unfold in a piece-meal fashion, (3) randomly, (4) via the genome, (5) lacking a specialized internal mechanism geared specifically toward change, and (6) implemented through the germ cells, (7) that are recombined, and (8) interact as equal participants.[5]

My personal perspective on the outlined problems of biological evolution runs as follows: (1) there are diverse mechanisms of change, (2) change is carried out piece-meal via minor isolated innovations as well as in a comprehensive sweeping fashion, (3) changes may be random or ordered, and (4) be implemented through the genome as well as the organism as a whole, (5) contain specialized internal structures to implement changes, (6) these mechanisms of change are incorporated into both somatic and germ cells, and the latter ones (7) are functionally distinct, and (8) interact as equals.

These issues will receive a detailed treatment in later chapters.

FIGURE 1. Combinations of different parameters defining the process of biological change.

| A | B | C | D | $E_1$ | | | | | | | | $E_2$ | | | | | | | |
|---|---|---|---|---|---|---|---|---|---|---|---|---|---|---|---|---|---|---|---|
| | | | | $F_1$ | | | | $F_2$ | | | | $F_1$ | | | | $F_2$ | | | |
| | | | | $G_1$ | | $G_2$ | | $G_1$ | | $G_2$ | | $G_1$ | | $G_2$ | | $G_1$ | | $G_2$ | |
| | | | | $H_1$ | $H_2$ | $H_1$ | $H_2$ | $H_1$ | $H_2$ | $H_1$ | $H_2$ | $H_1$ | $H_2$ | $H_1$ | $H_2$ | $H_1$ | $H_2$ | $H_1$ | $H_2$ |
| 1 | 1 | 1 | 1 | | | | | | | | | | | | | | | | |
| | | | 2 | | | | | | | | | | | | | | | | |
| | | 2 | 1 | | | | | | | | | | | | | | | | |
| | | | 2 | | | | | | | | | | | | | | | | |
| | 2 | 1 | 1 | | | | | | | | | | | | | | | | |
| | | | 2 | | | | | | | | | | | | | | | | |
| | | 2 | 1 | | | | | | | | | | | | | | | | |
| | | | 2 | | | | | | | | | | | | | | | | |
| 2 | 1 | 1 | 1 | | | | | | | | | | | | | | | | |
| | | | 2 | | | | | | | | | | | | | | | | |
| | | 2 | 1 | | | | | | | | | | | | | | | | |
| | | | 2 | | | | | | | | | | | | | | | | |
| | 2 | 1 | 1 | | | | | | | | | | | | | | | | |
| | | | 2 | | | | | | | | | | | | | | | | |
| | | 2 | 1 | | | | | | | | | | | | | | | | |
| | | | 2 | | | | | | | | | | | | | | | | |

## 3. A SYSTEM'S VIEW ON THE BOOK

I believe that the best way to summarize the present work is from the *systems* perspective. In keeping to the systems approach, any problem involves at least five dimensions: function, structure, process, operator, and genesis.

The development of this approach is heavily indebted to Jamshid Gharajedaghi.[16] His analysis of a system incorporates three dimensions - function, structure, and operation. I have added two more dimensions to this list - operator and genesis. As a result the system to be analyzed is subjected to the following dissection: the purpose or what final product will actually implement a given objective – *function;* ordering the elements into a totality – *structure;* transformation of primary ingredients into the final product - *process;* tools used to implement the process - *operator;* and the impact of the prehistory of the system upon all the above variables - *genesis*. In principle, all these variables are *independent*.

The majority of scholars, however, attempt to find one single dimension which determines all others. This is typical of functionalism, structuralism, operationalism, etc. Unnecessary arguments among scholars in the same field frequently arise as a result of some system being viewed from a single narrow perspective.

Using the above mentioned multidimendional methodology, the present book, from the *functional* point of view, represents an attempt to show that each stage of the evolutionary process - change, selection, and heredity - incorporates a multitude of interwoven mechanisms that differ in the basic principles underlying their operation. My focus is upon the possible somatic mechanism of change and cancer as its pathological manifestation.

This kind of differentiation/integration analysis of evolution is rather foreign to biological science. While evolutionary theory has undergone change over time, each historic period has advanced its own reigning doctrine which proclaims one single universal in time and in space principle that governs the process of evolution. The currently prevailing doctrine is the neo-Darwinist theory of evolution, which takes the following fundamental principles for granted: change is random, i.e., based on chance; selection is rooted in the struggle for survival; and heredity is encapsulated in the genome. These basic principles are untouchable and are not subject to evolutionary development. Certainly, there is some marginal, but extensive, research spearheading new approaches to biological evolution outside the traditional framework. These approaches, however, also tend to proclaim a single governing principle of evolution as being universal in time and space.

From the *structural* point of view, the book is divided into three parts with some parts preceded by an Introduction and followed by a Conclusion. Part One presents the analogy between societal and biological systems. It is more common for social sciences to appropriate ideas from the natural sciences, especially from physics, engineering, and biology. There are well-known and wide-ranging analogies between society and a machine, an organism, or a biological colony. While the reverse analogies are infrequent, they could be quite informative. One well-known example is Darwin's use of the idea of the struggle for survival, borrowed from Malthus's treatise on political economy. I know of only a single work written by a Russian-Ukrainian scholar, Michael Tugan-Baranovskyi (1865-1919),[17] which examines in a *systematic* fashion the impact of the ideas in social sciences on the natural sciences.

I have employed concepts derived from social systems in my analysis of both the macro and the microlevel of a biological system. By spotlighting the R&D sector as an intrinsic component of the economy's overall structure, we can draw interesting parallels with biological systems which might also incorporate an analogous internal, ordered, or, more precisely, semiordered, sector of change. Our nebulous analogy becomes more concrete when we examine the multi-stage process of R&D and the diverse mechanisms including the mechanisms of random mutations) associated each stage.

My approach toward the definition of a sex and the feasibility of multi-sexual reproduction was suggested by an analogy to the division of powers. Analogies to the socio-political systems at the microlevel are more relevant in the analysis of cells.[6] Just as there are social deviants, there are cells which either induce a positive change in the organism or cause its demise. The reader may have surmised by now that I regard cancer cells as deviant type cells that might attempt a drastic and rapid restructuring of an organism and be particularly successful when the organism is under stress and its immune systems is weak.

Part Two examines the general characteristics of two plausible normal mechanisms of change based on somatic and/or germ cells. It also explores specific characteristics of the internal mechanism of change and the related topic of the emergence of sexes over the course of evolution.

Finally, in Part Three, I elaborate upon my hypothesis that cancer is a pathological manifestation of the somatic mechanism of change. The somatic mechanism of change itself has been poorly explored. Therefore, my discussion is based primarily on the well-documented pathologies of this mechanism. The sequence of analysis from the pathology (cancer) to the normal mechanism of somatic change was turned down in order to make my analysis more coherent. A reader who may be unnerved by my approach can return to the chapter on the normal operation of the somatic mechanism of change after reading about its pathological expressions.

In the Conclusion to the book, I have tried to bring together all the individual hypotheses which I have advanced in the course of analyzing my main hypothesis.

My other aim in the Conclusion is to provide some justification for my intervention into the complex world of biology. Perhaps the arguments expounded in this section might be of interest independent of the actual discipline explored in the present work.

The *process* oriented aspect of the present book is closely tied to its methodology. The crux of my methodology in probing into the mechanisms of biological change is the systems approach. I have borrowed the idea of finding isomorphic structures shared by different systems and expressed rather loosely via analogies. I spotlight the parallels between biological and social, especially economic, systems, but I also embrace such fields as origami, music, etc. I have also made extensive use of other methodological devices, namely the multidimensional (function, structure, process, operator, and genesis treated as independent variables) analysis of the problem and the integration of the various facets into a holistic structure. The currently prevailing trends tend to reduce the whole to a single dimension regarding other dimensions as dependent variables.

As multi-faced as I have tried to be in my presentation, my focus on the single issue of the mechanism of change in biological evolution, in fact, the very controversial somatic mechanism, is not without certain flaws. The reader might get the impression that I am trying to enthrone *one decisive factor* governing biological evolution and subordinate the development of the living creatures to this one factor. I realize the pitfalls of such a one-factor methodology. Examples of Freudism, Marxism, etc. reveal the harm that can be inflicted by such a narrow approach. However, my focus upon this unique factor skewed my analysis of biological evolution, and I hope the reader will be understanding.

In my analysis of the biological system, I have also drawn upon certain philosophical categories, and I have elaborated upon those categories associated primarily with indeterminism. It is in connection with the category of indeterminism that I consider the idea of a tunnel process, which seems highly pertinent to biological systems. Essentially, the tunnel process connotes the idea that development is a two-way process: from the end, i.e., adaptation to demands imposed by the environment, as well as from the beginning, i.e., internal restructuring of the genetic make-up aimed at creating the potential. Eventually, over the course of many unforeseen events, this potential can give rise to structures which will respond to the demands imposed by the environment.

I further believe that natural philosophy can be a powerful methodological tool if used at the appropriate stages of the investigation of new problems. Generally speaking, natural philosophy connotes speculative meditations on natural phenomena. Based upon certain facts, speculations of this sort may not require direct experimental verification. History of science, on the other hand, has demonstrated the virtues of the experimental method in molding new scientific theories so most scholars feel a deep-rooted distrust for natural philosophy.

A number of reputable scholars are also vying to rehabilitate natural philosophy. For example, Ernst Meyer acknowledges the crucial role of natural philosophy in preserving and developing the concept of a complex organic world during the reign of Newton's mechanistic reductionist methodology.[18]

It seems to me that natural philosophy fertilized by contemporary ideas on systems' organization (ideas born of specific disciplines or, indirectly, via the general systems theory) can be very helpful in the analysis of natural phenomena. Natural philosophy provides Weltanschauung, or a broad perspective, into the essence of the problem. It becomes indispensable at the *initial* stages when the foundation for a new concept is being laid and the most fundamental issues in the respective field are being explored. Under these circumstances, we must start with certain axioms, meaning certain assumptions must be taken on *faith* Here, natural philosophy helps steer us in our choice of axioms and subsequent course of research. And, since a developed society allows for multiple philosophical paradigms to coexist, different axioms picked by different scholars preserve the pluralism in respective field and help prevent a false course of research from monopolizing all resources.

As the new theory takes shape, there is a need to formulate more mature hypotheses that will be ultimately amenable to formal mathematical representation and experimentation. The issue, therefore, is not some abstract validity of natural philosophy, but the stage in the investigation of a specific problem at which it ought not to be discarded. To repeat, I believe that natural philosophy is helpful in dealing with the early stages of completely novel investigations. The more a given field is explored and the more a given hypothesis follows a trodded path, the less need there is for philosophical speculations. Moreover, the use of natural philosophy, especially its early stages, is fraught with superficial vulgarization if one overextends its use to analyze specific facts and thus bypasses certain crucial intermediate stages such as the formulation of experimentally verifiable hypotheses.

In the present context, the *operator* is the book's author. I have made the reader familiar with the author's person in order to make the point that, in spite of the book's lofty subject matter – cancer and the mechanisms of biological change – it is written by someone who, while not a specialist on the subject, is not a charlatan.

Finally, the *genesis* of the book - the history of how it was written. This aspect reflects the *source material* and *preliminary publications* of my hypothesis regarding the mechanisms of change.

Much of the material I have used in the book is concerned with molecular biology and the various stages of cancer, and was published in professional books and journals as well as in the Science, Nature, Science Times section of The New York Times, etc. I recognized the need to discuss my ideas with professional biologists before they embarked upon their own publications. I did just that gaining much insight from many valuable discussions with biologists. Certainly, the people with whom I discussed my ideas cannot be accused of unrestrained admiration for my views. I would venture to say, however, that the majority of my colleagues did not consider my views to be *generally known*, *trivial*, or *outright wrong*. Most people with whom I have discussed my meditations on biology have opted to postpone their final verdicts until some experimental results are in.

An abridged version of the paper containing my biological hypothesis appeared in 1991 as an article in an established Western journal.[19] At first, I shared my raw ideas with my friends, mostly specialists in biology. At that time, I did not feel my ideas were publishable. The first published statement came in 1984, covering less than a page of my book[20] (pp. 93-94) devoted to new trends in systems theory. A more elaborate version of my deliberations was published in the proceedings of an International Conference.[21] The final published report encompassed three pages of rather terse text.[22] (pp. 44-46). I was also honored when a former Soviet journalist, now living the United States, expressed interest in my biological speculations. Mark Popovsky – the author of many and articles on biology and medicine (some came out in Samizdat), – published a large article entitled "At the Crossroads of Science, or Cancer Through the Eyes of an Economist." The article appeared in the November 2, 1990 issue of the American Russian-language daily, Novoye Russkoe Slovo.

The book is intended for a broad spectrum of biologists, medical professionals, systems scholars, and philosophers. There is also another group of researchers who stands to gain from this book – those involved in genetic programming. By imitating the process of natural selection genetic programming seeks to solve problems without specifying exact procedures.

Unfortunately, this new discipline still employs rather rudimentary notions about the mechanisms of biological change (random mutations and crossing without taking into account the specific functions fulfilled by the crossing organisms).[7]

At the same time, genetic programming is actually not limited to uncovering and implementing nature-given evolutionary mechanisms. It creates room for hypotheses regarding novel forms of reproduction and tests them by simulation. Since these new types of mechanisms have not yet been found in nature, the mere feasibility of their existence is heuristically conducive to our search for them in nature.

I should be most grateful to all readers who take the time to send me their comments regarding my deliberations.

## NOTES TO INRODUCTION

[1]. This approach is reminiscent of Oscar Wilde's aphorism that art is not a reflection of life, but life is a reflection of art.
[2]. The following statement regarding the theory cancer made by an outstanding Russian pathologist Ippolit Davydovskii is still highly pertinent: "No other branch of medical science can rival oncology in exposing the glaring up between the wealth of factual data and chaos of theoretcial explanations of the etiology and nature of tumors. This abundance of facts and theories have not fostered a real theory of cancer."[23] p. 16.
[3]. As James Graham mentioned: "I wrote this book to convince all who read it that cancer played a major role in evolution and in doing that I say nothing negative about the disease. At times I may even seem enthusiastic about its function or, at the very least, its results. This does not mean that I think it is doing anything "good" when it causes suffering and death. Although I am convinced that cancer is a biological function and that we would not exist without it, that is no reason to decrease efforts to cure anyone afflicted with it or to eliminate it. In fact, although I did not write the book as a guide to cancer researchers, it is conceivable (although I think it extremely remote) that someone in that most demanding field may find in its pages the inspiration to undertake a new approach that will prove fruitful. Nothing would please me more."[24] p. xiii.
[4]. "Monograph - a way to present a fanatical idea" - this definition is given in "Under the Mobious Strip" - a humor-filled book published in 1993 by a researcher at the Central Economic- Mathematical Institute of the Russian Academy of Sciences. This definition suits the present opus perfectly.
[5]. As James Shapiro mentioned,"...the prevailing evolutionary wisdom [is] based on notions of piece-meal, stochastic genetic change due to replication errors and physico-chemical instabilities."[25]

[6]. "The body of an animal can be viewed as a society or ecosystem whose individual members are cells, reproducing by cell division and organized into collaborative assemblies or tissues."[26] p.1187.

[7]. "To accomplish it, genetic programming starts with a primordial ooze of randomly generated computer programs composed of the available programmatic ingredients, and breeds the population using the Darwinian principle of survival of the fittest and an analog of the naturally occurring genetic operation of crossover (sexual recombination)."[27] (p.1)

*PART ONE:*

## AN ANALOGY BETWEEN SOCIO-ECONOMIC AND BIOLOGICAL MECHANISMS OF CHANGE

My aim in this part is to prepare the reader for subsequent direct parallels, bypassing any intermediate language, between societal and biological systems. In other words, generalized examples of the mechanisms of change that are incorporated into societal systems reveal the importance of subsequent analogies with the process of change taking place in the biological world.
Let us begin with the economic system.

# CHAPTER 1

# ECONOMIC MECHANISMS OF CHANGE AND BIOLOGICAL EVOLUTION

## 1. GENERAL FEATURES

### 1.1. Paradigms of Economic and Biological Evolution

Theories of economic development and their influences upon biological thought have a long and distinguished history.

On September 28, 1838, Charles Darwin picked up Thomas Malthus's book, Essay on the Principle of Population. This work had a decisive influence upon the development of Darwin's concept of evolution. The economic paradigm, which Darwin borrowed from Malthus, led him to formulate the *survival of the fittest* principle as the cornerstone of his evolutionary theory.

Apart from incorporating the idea of natural selection advanced by an economist, the post-Darwinian concept of evolution implicitly reflects the prevailing economic paradigm of the time. Let me clarify this point.

Economists in the 19th century believed that *change* – technological innovations – was primarily the result of random processes engaged in by individual inventors. Consequently, the attention of scholars was centered on the mechanism of *selection*, namely the market where competition determined (selected) the fittest - the ones most adapted to the environment. The products selected received reinforcement from the institutions whose existence was a given within the general dynamics of the model - *heredity*.

Darwin's theory presumes that biological changes are random, that selection amounts to the survival of the fittest, and that the entire organism helps preserve (pass on) the hereditary traits.

Now, economic science has witnessed a profound change in its understanding of both the interaction among change, selection and preservation of innovations, as well as each of these processes individually.

Modern economic science possesses sophisticated tools to deal not only with the competitive win-lose paradigm of selection, but also with cooperative win-win relationships. These mixed competitive-cooperative market processes are further enhanced by both endogenous and exogenous coordinating bodies, such as the stock market and the government. Economic science has also made great strides in understanding the value of preserving and nurturing corporate culture ("genetic code") that has proved its worth in the world of business.

Rapid technological progress in the second half of this century has brought about a deep transformation in the paradigms of economic change. The scale of change and its more structured output have given rise to a new sector (sphere) of Research and Development (R&D in the most general sense of the word) that is incorporated in the economy as a whole. The performance of this sector is far from random, something that is especially evident in applied science and engineering, although many minor improvements are somewhat random, coming from factory workers or similar isolated sources.

Economic theories of scientific/technological change are at the embryonic stage compared to the rather evolved theories of production, distribution, organization and policy implementation.

> "The central problem facing England, France, and the United States was one of innovation, of developing new techniques and technologies, and of diffusion of these recently developed technologies throughout the economy. The "economics" of this process of innovation, though discussed by Schumpeter, Hicks, and Robinson, among others, is not as well understood as that of the production and allocation of goods and services using established techniques. Only recently has it become once again the center of attention."[28] (p. 4.)

Still, work on R&D gives us a glimpse of the mechanisms of economic change and their interactions with the mechanism of selection and preservation of hereditary traits. At this time, I am interested in tracing the truly revolutionary influence exerted by the R&D sector upon the conventional schemes of economic development, i.e., the emergence of an internal (endogenous) mechanism of change aimed at semi-ordered development of innovations, which does not preclude completely random processes. What makes this analogy relevant to biological issues is the fact that the theory of evolution is still very much enthralled by the assumption that biological changes occur via random mutations - an assumption which shifts the focus to the mechanism of selection and subsequent reinforcement of selected genetic characteristics.

## 1.2. Change, Selection, and Heredity

There are three basic processes that shape evolution: *change* (formation of a manifold), *selection*, and *heredity*. The implementation of these processes, both individually and as a set, is ordered to a varying degree, and is exemplified by the economic systems.

Contributing to the synthesis of these processes is their mutual feedback, i.e., there is a certain order governing the interaction among these processes. Whereas previously, within the framework of the traditional paradigm of economic development, the creation of the manifold and its realization were largely disassociated, the new paradigm, which includes the R&D sector as an integral part of the economic system, ties these two processes together. For instance, this interplay is manifest in the fact that the market, being a mechanism of realization of the manifold, not only imposes demands upon the manifold, but is also one of its major financing sources. Each of the three processes mentioned above also entails diverse mechanisms of non-random operations; a number of mechanisms related to the R&D sector will be examined below. Therefore, while random mechanisms have not vanished they play a subordinate role.

In analyzing a system's dynamics, we have many ways of assigning priorities to such processes as the formation of a manifold, its implementation and its reinforcement. For instance, if implementation is deemed most important, the other two processes - change and reinforcement - become subordinate to that goal. If, on the other hand, the key aim is the expansion of the manifold, then implementation and reinforcement must be made to support expansion.

Economic science has typically operated within the following conceptual framework: scientific/technological change (the manifold) and the preservation of accumulated experience (reinforcement) are subordinated to the attainment of optimality (equilibrium) via a coordinated interaction of the participants (selection). This approach is particularly prominent in optimization models with endogenous technological progress.[29] These models implicitly assume the existence of a planning mechanism which performs an orderly and efficient search/selection for the best allocation of resources employed in the production of existing products by using existing technologies as well as the development of new products and new technologies.

The ideas of Joseph Schumpeter surpassed classical economic theory by emphasizing the role of change, i.e., emergence of economic disequilibrium brought about by the random actions of entrepreneurs. The mechanism of coordination, which is supposed to balance the economy and prevent its collapse, was relegated to a subordinate role.[30] The supporting function of the equilibrium and stabilization mechanisms is two-fold: first, not to impede the expansion of the manifold, and, second, to prevent chaotic disintegration of the system.

Generally speaking, I agree with Schumpeter's approach to economic development. However, I think he has overstated the role of spontaneous events in driving economic innovation. The distinction between the process of development and the process of implementation is blurred, i.e., he views technological progress primarily through its performance in the market, overlooking non-market horizontal institutions, e.g.,universities, foundations, or government institutions that intervene in the process of development. I shall address this issue of diverse institutions involved in non-random scientific/technological progress.

Biological concepts of evolution are still largely governed by the pre-Schumpeterian paradigm: the cornerstone of the biological sytem is the process of implementation and reinforcement severed from the process of change; change itself evolves in a random fashion similar to Schumpeter's scheme. The biological paradigm advanced here, which could be termed *integrative evolution*, presupposes a certain regularity in the process of change, selection and heredity (random elements persist, but are limited). My emphasis is the interaction among these processes via a feedabck mechanisms such as adaptive mutations in the interplay between change and selection.

## 1.3. Allocation of Resources Between Innovation and the Preservation of Existing Wealth

Whatever will be the solution to this conflict between development and stabilization of the economic system, in order to develop (survive), society has had to resolve the problem of the allocation of resources among: a) raising the standard of living and productivity level today; b) increasing output based primarily on the already existing technologies geared toward raising the standard of living in the near future; and c) investing in R&D, which might bear fruit tomorrow or the day after tomorrow or even in the more remote future.

Another factor in resource allocation is the service life of resources employed in both the established and the new methods of production. In terms of a simple dichotomy, we have durable goods, which require one-time expenses, e.g., equipment, buildings, and non-durable goods, which require day-to-day expenditures, e.g., materials, energy. The first group forms the skeleton of the system, its anatomy, and the second supports the ongoing (physiological) processes.

At the microeconomic level, the problem of resource allocation between today's needs and future development is resolved via the interests of the various individuals/groups who want to channel more resources into their own domain. All of the accusations against capitalists, such as their selfishness or exploitation of the workers, basically reflect the fact that any business struggling to develop (survive) tries to channel resources away from today's needs of the workers into expansion of production (which might call for investment in new ideas) in order to satisfy tomorrow's or even more remote needs. On the other hand, and this is especially true when the overall standard of living is low, the workers are more interested in satisfying their current needs. Businessmen, together with the creators of new ideas, regard workers, who merely carry out their ideas in a rather routine manner, as exploiters, because the latter take the bulk of the increment generated by new technological and organizational innovations.

This global view of the economy creates the problem of resource allocation among goods of different temporal significance. The problem is resolved through the exchange of goods among the many producers who pursue their own self-interests. Conflicts that arise are resolved either through the center (state) or through a struggle, within certain limits, among the various groups.

The problem of resource allocation between investment in innovation and current output has interesting repercussions for biological systems. For instance, it surfaced in connection with the so called *C-value paradox* and the related *selfish genes*. I believe that selfish genes are invovled in the mechanism of change inside the genome, and considerable resources diverted to support them represent the cell's expenditures on the internal "sector of R&D".

## 2. R&D AS A PROTRACTED SYSTEM

### 2.1. The Multi-Stage Process of Change and the Respective Mechanisms of Operation

Let us now examine, stage by stage, the structure of the R&D sector and the diversity of the mechanisms of operation arising as a result of some fundamental differences among these stages.

General ideas associated with fundamental science are the fruits of basic research. These ideas are subsequently transformed into more formal concepts addressed by applied science, and then evolve into engineering designs subject to trial by semi-industrial pilot projects which might eventually develop into technologies of mass production. There are crucial differences among the various stages of development and implementation of new ideas which lead us to the notion of a *multi-stage process* of development of an economic system. This sort of linear stage by stage description does not imply any rigid order in the implementation of each stage. The various stages can evolve concurrently with feedback. Even the last of the above mentioned stages is not final because practical experience may point to further improvement.

Each stage of the R&D process has its own peculiar features and calls for an appropriate mechanism of operation. One essential precondition required by these mechanisms is the existence of horizontal relations, i.e., independence and parity of the various organizations taking part in the process of creation. Another crucial factor is pluralism, meaning a multitude of independent organizations working in parallel and financed by different sources. Vertical mechanisms, where some units are subordinated command-like to other units, that are manifest in the various forms of government intervention, e.g., tax breaks for corporations, funding for university research, direct funding, should not be ignored.

Horizontal mechanisms can be classified into market or non-market types. At the initial stages of development characterized by the birth of new ideas, the mechanism of selection ought to rely on the producer rather than some outside institution. The more innovative an idea, the more its originator should be shielded from professionals in the respective field. First of all, specialists are frequently *unable* to evaluate a radically new idea if only because it forces them to revamp the entrenched mode of thinking and enter an entirely new and still rather nebulous scheme proposed by the innovator. Secondly, it is not uncommon for specialists not to want to recognize new ideas in their field because it might hurt their professional status or detract from some other valuable asset they possess.

As new ideas evolve, the role of external mechanisms of selection increases, especially expert opinion, but not the market which typically handles the sale of the fruits of one's labor. The diversity of organizations involved in this stage of R&D prevents any single one from monopolizing the evaluation procedure. The organizational framework characteristic of this stage is the institution of foundations.

In summary, since the initial stages of R&D require that the producer be shielded from the overly pragmatic criteria of the consumer, what are called for are different types of non-market mechanisms. Universities, independent inventors, foundations, etc. represent the various institutions suitable at this particular stage.

As abstract ideas begin to materialize into more worldly things and the consumer's role of selection increases, market instruments, ranging from venture capital to public corporations, become the dominant force.

The above considerations pertain to the one-time implementation of innovations (in the organizational sense as well, i.e., confined to a given firm). The subsequent evolution of innovation entails expansion, meaning the capturing of a new habitat. As the process of innovation evolves and branches out, it tends to transform itself into a more imitative endeavor. This leads us to a very interesting realm of economics pioneered by a prominent economist Joseph Schumpeter. The issues addressed by this field include the alliance between R&D and the production sectors, the designation of organizations best suited to be innovators or imitators, the ratio of innovators to imitators, etc.

Market and non-market mechanisms in the R&D sphere are not isolated from each other. R&D conducted at corporations is coordinated not only via the market, for example, joint ventures set up for specific projects, but might involve institutions belonging to non-market horizontal mechanisms: organization of conferences, special journals and publications, etc.

Western countries have demonstrated the viability of diverse institutions by overcoming the notion that the R&D sphere can operate only if governed by one of two extremes, either the market or the state. These countries have (and continue to search for) a multitude of mechanisms designed for the creation of new things. The mechanisms work by integrating vertical structures, e.g., government intervention,with horizontal market and non-market structures. The more imitative ventures could remain, to a large extent, the prerogative of the market.

At the present time, biological science lacks this kind of a multi-stage analysis of the process of change if only because biology has long been dominated by the theory of random mechanisms of change. The search for internal mechanisms of change is in its infancy, still largely devoted to affirming the mere fact of its existence.

## 2.2. Tunnel Process

R&D can proceed from the *beginning* as well from the *end*. This two-ended approach, which I refer to as the *tunnel process*, is most conducive to rapid and effective progress in the R&D sector.

In the simplest case, an economic system can be represented by the production chain from consumer goods and services to military output. This means that a person takes from nature what he or she needs. Vestiges of this still survive – the collection of wild berries, mushrooms, and other gifts of nature.

As man evolved, or, perhaps, from the very beginning, there emerged multi-link production chains between nature and the final product. Increasingly sophisticated technologies, comprised of many stages with many intermediate goods, were developed before the final product could be obtained. These production chains are manifest primarily in the production of tools, including second-order tools, that is, tools for making other tools.

However, the production chains thus formed were largely atomized and relatively short. Once the tools and the intermediate products became more versatile, this tangle of production chains was transformed into a unified network.

As long as this network remained relatively simple, a change in any one of its nodes could be completely and consistently traced directly to the end result, the final product. Therefore, with respect to the deterministic domain of the economic system, i.e., the domain which yields consistent and complete links between some initial and future states, changes can proceed from the end. This means that the process of development is driven by a well defined practical objective, namely, the creation of new products or technologies which are integrated in the overall network in a rather complete and consistent manner. Under these conditions, the results are judged directly by the consumers.

At some stage of human evolution a question arose which marked a revolution in the history of the human knowledge: "Why not begin parallel development from the start?" The question was posed in the following way: "Why not proceed in a parallel fashion, taking any arbitrary link as a starting point and assuming it to be impossible to construct any local criterion that would allow that link to be integrated in a complete and consistent manner with the rest of the coordinated network?" Tremendous benefits could be reaped from this kind of a parallel development, akin to digging a canal from both ends. At the earliest stages of development, the network is comprised of units (links) that mainly produce versatile products. If these products can be developed quickly and ways of integrating them into the existing network found, this could accelerate the development of the system as a whole. These considerations give rise to basic and applied science, which eventually contributes to the creation of new products and technologies.

The tremendous benefits of the two-ended approach, however, exact a heavy toll. The problem is that at the beginning the potential worth of any given undertaking may be entirely unclear. In other words, the stage of development attained by individual units is not complete and consistent with the network as a whole. For this reason, investment in the initial link could easily lead nowhere or to a dead end.

Thus, many sectors of the economy become indeterministic due to the gaps arising as a result of development being initiated at the beginning; in effect, it is impossible to link the initial and the final states in a consistent and complete manner, even in terms of probabilities.[1]

With respect to biological evolution, tunnel process implies the existence of external as well as internal sources of change. It is further surmised that external sources of change operate from the *end*, meaning that they reflect external conditions (environment). Internal sources of change work from the beginning, creating predisposition toward future development. The joint action of these two sources of change creates the so called *tunnel process* in biological evolution. Possibly the genetic changes induced at the beginning are accumulated in the genome prior to their ultimate expression in the phenotype. Consequently, paleontologists need not seek all the intermediate changes, that might be redundant or even harmful, in the evolution of a phenotype of a new species.

## 2.3. Calculating the Impact of a Given State upon Its Subsequent Development at a Point of Discontinuity

The generalization of the idea of a tunnel process brings us directly to the problem and degree of indeterminism. Indeterminism is one of the least developed philosophical categories. One indirect confirmation of this statement is the fact that mainstream philosophical literature, including contemporary theoretical investigations by Karl Popper,[31] fails to evoke the notion of degree of indeterminism: As a rule, indeterminism is opposed to determinism, with no mention of any intermediate stages.

We know that the concept of *measure* played a crucial role in the history of human knowledge. At the earliest stages, man operated with binary relationships reflecting extremes, such as hot-cold, black-white, etc. This is evident in a number of ancient languages. Subsequently, mankind developed the notion of *phases* reflecting the measure ascribed to a certain state – for instance, torrid-hot-warm-cool-cold-freezing. At a still more advanced level, the notion of temperature was introduced, allowing the user to measure change with the required degree of precision.

The notion of a phase is crucial in two respects: first, in measuring a given parameter with the required degree of precision, and, second, in developing methods to operate with structures associated with a given phase. For example, different phases of matter - plasma, gas, liquid, solid, etc. require different technologies.

In summing up my approach toward the category of indeterminism, I have arrived at a general systems concept of a multi-stage (multi-phase) process of change and the various mechanisms used to implement it.

I have (or, so I believe) uncovered a new and very important phase. Independently and based on different precepts, a number of scholars from the mathematical community have expressed interest in this philosophical category. Actually, one mathematician, Ron Atkin has provided a conceptual framework for this idea.[32] The phase I have uncovered within the spectrum of indeterminism involves the creation of *predisposition* toward development. Predisposition is a state which, on the one hand, is not completely ordered, not even in terms of probabilities, but neither is it complete jumble (unlike chaos, which, according to modern science, does exhibit certain patterns[33]).

In order to alleviate the potential waste caused by the gaps in the multistage process of change, it is essential to be able to evaluate the states belonging to these points of discontinuity, gaps, in terms of their potential impact on the future of the system. I believe this evaluation is rooted in the aesthetic method, i.e., measuring predisposition via the degree of *beauty*.[34]

The aesthetic method applies to situations where the path from the current state to some future desirable state cannot be fully and consistently calculated. These conditions call for the creation of the potential. Potential represents system's predisposition toward development. It is different from the probabilities-based approach which extrapolates on the basis of previous frequencies the impact of the current state upon the future. This concept of the potential is helpful in evaluating *unique* situations, since it does not require any knowledge of probability distributions in analogous cases in the past.

The system's potential creates structures aimed at:

1) inducing the environment in which the system is immersed to the system's advantage; 2) preparing the system to turn unexpected outcomes to the system's advantage, and 3) absorbing or reducing the shocks of unexpected events harmful to the system.

The creation of potential is based on the following four elements:

1) essential parameters as independent variables, 2) unconditional valuations of essential parameters, 3) relational parameters as independent variables, and 4) unconditional valuations of relational parameters.

These four elements make up the so-called weight function which measures the power of the potential as a scalar quantity.

What is new about this approach is that it introduces, on the one hand, unconditional valuations for essential parameters, and, on the other, relational parameters as independent variables. It is precisely the incompleteness/ inconsistency of the links between the present and the future states that necessitates the introduction of unconditional valuations of essential parameters supplemented by appropriate valuations of relational parameters. As part of this scheme the relational parameters also become guiding parameters on a par with the essential ones.

Uncovering the constituent elements comprising the potential, as well as the methods of these elements' synthesis, is really a rather ambitious attempt on my part to demonstrate that the category of beauty can be represented structurally, i.e., it can be decomposed, evaluated, and then synthesized.

Since scientific discoveries and inventions fall into the already existing R&D network, one should look at the system's predisposition toward innovation rather than focus on the random events leading to some particular novelty.

The idea of measuring predispositions by means of the aesthetic method extends beyond the human domain. It seems that the complex web of interactions among cells within an organism, among living creatures, and even chemical[35] and physical forces, is predicated not only upon the principles of determinism, but also upon the creation of predispositions.

## 2.4. The Organizational Structure of the R&D Sector and the Variety of Methods for Supporting Its Operation

The various stages of R&D correspond to different organizational structures. The R&D sector is comprised of many diverse institutions, ranging from universities, which deal primarily with fundamental principles underlying innovation, to firms, which combine applied research with the development of new technologies and their implementation, to inventors who, for the most part, develop new technologies based on the already known principles of pure and applied science.

In terms of power, the R&D sector is comprised primarily of *powerful* research organizations such as universities, specialized scientific institutes, e.g., laboratories, design groups, and special divisions within corporations. This sector also includes individual (or small group) *garage* inventors, individuals specializing in innovation within the corporation, etc. Between these large-scale organizations and "small-time" inventors there is an entire spectrum of organizations of varying power, including firms specializing in some narrow area of research, small groups involved in the development of new technologies in mid-size companies, etc. The R&D sector is not a homogeneous entity. Apart from the specialized and organizationally distinct entities involved specifically in innovation, this sector includes subdivisions dealing with specific products as well as autonomous individual-inventors dissolved in the economic system.

This heterogeneous structure of the R&D sector is financed through a wide array of sources that can be divided into several categories. The first is the revenue generated by a given unit from the sale of the fruits of its research, with money being used to finance new research. The second type of source is the unit's own money received from the sale of some final product/service, not the fruits of its research, or money saved from previous operations. The third source of financing is the government, i.e., the system as a whole. The fourth type includes donations from individuals/organizations directly or indirectly via money from private funds, inherited wealth, and so on. The fifth type of source is bank credit.

# ECONOMIC AND BIOLOGICAL MECHANISMS OF CHANGE

All of these sources for financing innovations are intertwined. This diversity creates more room for speculative pioneering research. It also raises a two-fold problem of the innovator's autonomy.

The first issue is the complexity of the organizational framework established by the innovator. Implementation of major innovations by such complex organizations as corporations is made possible, to a large extent, by profit. The capital raised through the sale of stock (this represents capital accumulated by individuals from their previous work) and the use of previously generated profit allows corporations to finance innovative ventures that are beyond the risk threshold tolerated by banks or even stock holders; bank credit is used by corporations to finance safer projects. Individual inventors, on the other hand, gain independence by setting up a relatively simple self-financing organizational environment. This drastically cuts down on current expenditures and permits reallocation of resources saved into innovative ventures to be recovered in the future.

The second facet of the problem is financing R&D units whose output is not for sale (which rules out profit-driven financing). This situation arises at the initial stages of basic research, and the institutions that come to the rescue under these circumstances include government, charities, etc.

It would be interesting to analyze the degree of autonomy exercised by the cells that partake of innovation in the light of the ideas expounded above and my assumption of an internal mechanism of change. A cell's autonomy may be expressed in a drastic alteration of its metabolism, e.g., a switch to a more primitive anaerobic metabolism that does not require oxygen), or in reduction in the number of receptors linking it to other cells, etc.

The more complex the R&D and productions sectors are, the more difficult it is to coordinate them and the more painful are the disorders in the R&D sector. Take, for instance, flaws in the engineering of nuclear power plants. When selection and reinforcement are carried out by more primitive centralized totalitarian regimes, the damage, which might extend far beyond the borders of a given country, can be immeasurable. It is not necessary to dwell on the dangers for mankind and our environment posed by genetic engineering or similar intrusions into the most basic building blocks of physical, chemical, or biological objects.[2]

## 3. THE LESSONS FROM THE ECONOMIC MECHANISMS OF CHANGE FOR THEIR BIOLOGICAL COUNTERPART

### 3.1. General Devices

What general devices relevant to biological evolution and biological mechanisms of change are suggested by our analysis of an economic system? It seems to me that one of these is program hierarchy: *program-changing programs*. Having discussed the various methods of operation associated with each stage of R&D, I would now like to consider these stages from a different perspective, namely in terms of a hierarchy of programs for creating innovation.

The category of a program as an independent object is a latecomer to the scientific scene. Its development was propelled by the great strides made by science in the twentieth century. The invention of computers, the study of the genetic code at the molecular level, the technology of mass production of complex goods – all these areas require a rigorous, sometimes even formalized, determination of complete and consistent relations among the various elements of the mechanism – precisely what the category of a program entails.

A behavioral program, which, according to Herbert Simon, governs the behavior of the system directly, will be denoted as a zero-level program. A first-level program changes the zero-level program. Changes in the zero-level program may be due to new information gathered over the course of its implementation or to the latent possibilities embodied in the first-level program. A second-level program works similarly upon the first level program. Moreover, assuming isomorphism among the various levels, the possibility of a feedback should not be ruled out. For instance, a third-level program can bring about changes in the second-level program (because of new information accumulated over the course of its execution or because of the possibilities intrinsic in the third-level program itself), or it can, itself, be altered because of changes in the second-level program.

The described hierarchy of programs resembles the process of change. Let us say that the zero-level program is a program which governs the production schedule for a particular product. Then, the first-level program is a program which generates the schedule. This program is based directly on the established principles of industrial management. Applied science represents a second-level program because it elaborates new principles of industrial organization. Basic science represents a third-level program, since it is supposed to alter, in a radical manner, the principles underlying the development of the principles of industrial organization.

As systems evolve and become more complex, the number of levels in a program hierarchy seems to increase. Perhaps living creatures, with the exception of man, possess about a three-tier hierarchy (behavioral program, a program that shapes it, and a program that changes the latter program, i.e., a learning program). In human beings the number may reach four, and in man-made artificial systems it may be even higher.

## 3.2. Specific Devices

What specific devices relevant to biological mechanisms of change are suggested by our analysis of an economic system?

The following ideas come to mind:

1) The production of goods needed today and based on routine procedures, and the creation of innovations needed tomorrow or in the more remote future, should be not viewed as isolated processes. At the microlevel, the actual volume of production of both kinds of products is determined by a tug-of-war between their respective producers who pursue their own interests. I shall qualify this statement later. There are many ways to resolve these conflicts - from the state exercising absolute control to the free struggle among the participants.

2) The creation of a scientific/technological idea is a multi-stage process of R&D with subsequent development of the final product and its realization by means of different mechanisms.

3) The multi-stage process of creation can give rise to the tunnel process, i.e., where the process of creation can be two-ended, originating at the beginning as well as from the end.

4) Gaps may arise within this multi-stage process of creation; evaluation of these gaps is done through the creation of predispositions.

5) The process of change can be viewed as a program hierarchy which includes program-changing programs.

6) In terms of its organization, the R&D sector incorporates a wide range of structures (units) supporting innovation - structures ranging from autonomous organizations, e.g., universities, foundations, to isolated individual inventors dissolved in the economic system and involved in innovative modifications of known technologies.

7) Innovation is supported through a variety of sources corresponding to the various organizational structures in the R&D sector.

Particularly important are those sources that ensure independence of organizations involved in the fundamental early stages of innovation.

8) Both innovators and imitators seek to elude the control exercised by market-driven interaction among economic units. However, while the innovator strives to gain greater autonomy, innovation propagated by the imitators tends to master the established network.

9) The more complex the R&D and productions sectors are, the more difficult it is to coordinate them and the more painful are the disorders in the R&D sector.

The above considerations on the relevance and instructiveness of economic ideas for the analysis of biological systems suggest the following hypotheses regarding the latter:

Hypothesis number one. Beside the many mechanisms of change based on external factors affecting the zero-level program in the genome, (radiation, chemicals, gene recombination), there exists a variety of internal mechanisms of change. The scope of change produced by these mechanisms ranges from minor changes to the birth of new species.

Hypothesis number two. The structure of the mechanisms of change within the genome can perhaps be analyzed in terms of program hierarchy.

Hypothesis number three. The role of the tunnel process in evolution, meaning that biological changes can proceed both from the end as well as from the beginning.

Hypothesis number four. Mechanisms of change, especially those which operate from the beginning, possess certain structures which are first introduced in the somatic cells with information subsequently transferred to the germ cells.

Hypothesis number five. When change is initiated at the beginning, there exist structures within the genome that create potentials capable of developing in many different directions.

Hypothesis number six. Minor changes originate primarily at the end, while major ones are initiated at the beginning.

Hypothesis number seven. With change initiated at the beginning, there exist structures in the genome which minimize the occurrence of dead-end mutations .

Hypothesis number eight. As an organism develops, it faces the problem of resource allocation between a) supporting the creation and operation of the genome with an internal mechanism of change, and b) speedier multiplication and growth of the cells governed by routine genetic programs.

Hypothesis number nine. The humoral regulatory system incorporates mechanisms that allow innovator cells to break away from the organism's control, to gain autonomy, particularly at the early stages of their development. Once having evolved into a critical mass, innovator cells may actually get rid of the old regulatory mechanism and take over the control of the organism.

## NOTES TO CHAPTER 1

[1]. The entire history of mathematics has exhibited two-ended development. One end was the need to solve practical problems, land surveying in particular. The other was the desire to explain the harmony of numbers as such. In Egypt the development of mathematics was driven by practical needs, such as geodesy and pyramid construction. Greek mathematics, Euclid and especially Pythagoras and his followers, professed aloofness from reality and the purity of mental constructs. Pythagoreans espousing the credo "numbers rule the world" despised practical applications of any kind. Only much later in the nineteenth and twentieth centuries, when the depths of mathematical structures were penetrated more deeply, did the two approaches begin to cross-fertilize each other, for instance, number theory, analysis, and theory of probability. Naturally, the connection between these two approaches hasnot been fully explained.

Originally, painting developed predominantly as a representation of the concrete. Gradually, as artistic sensibility evolved, the artists began to move away from concrete elements in their work. This is manifest, for instance, in the evolution of Vincent Van Gogh's painting comparing his *Boots with Laces* with *The Abandoned Quarry*. My real point is that, along with the painting of concrete objects, there also appeared painting based on abstract elements and colors. Although all great artists recognized abstract qualities, for a long time they were considered subordinate. Apparently only in the twentieth century, beginning with the work of Wassily Kandinsky, did abstract painting begin to assert its independence. All these developments have led to many different forms of synthesis of abstract and concrete structures, to the mutual benefit of both. For example, in a number of paintings by Michail Shemiakin, a delightful synthesis — using the motif of Russian folklore — is achieved between the concrete structures of Hieronymas Bosch and the abstract structures of Kandinsky.

[2]. Not surprisingly, from antiquity people feared "the tree of knowledge" or "the Pandora's Box" and certain advanced civilizations, such as India and China, chose to curb technological development for many centuries. The common thread running through the two myths - the Tree of Knowledge and the Pandora's box - was pointed out to me by the late Professor Boris Moishezon. I refer an interested reader to my article[36] which discusses the relationship between introvert and extrovert cultures.

CHAPTER 2

# MECHANISMS OF CHANGE IN SOCIO-POLITICAL SYSTEMS AND BIOLOGICAL EVOLUTION

Our discussion of socio-political systems is divided into two parts. I hope that the parallels between the socio-political and biological structures will help the reader understand not only the normal process of change, but the abnormal processes taking place in biological systems.

## 1. MACROLEVEL

What is called democracy seems to be a multidimensional, nonlinear entity. Its level of development is a function of many independent variables capable of assuming a range of different values. What is this multidimensional society? With respect to political systems, it incorporates such aspects as pluralism, not to be confused with relativism, democracy proper (unlike ochlocracy, or power of the mob), separation of power, and openness. Let us take a closer look at certain elements of a democratic society.

### 1.1. Separation of Power

At first glance, it seems that an enlightened monarchy, with one highly educated individual endowed with unlimited power and long-term commitment, should be capable of creating a new social institution quickly and efficiently with the best people in the country acting as the monarch's advisors. However, the "crazy" Charles Montesquieu (1689-1755) tried to show that representative democracy, with separation of power based on the principle of *checks and balances*, is, quite paradoxically, more conducive to a country's development. Under the system he proposed, the creation of new social institutions might take more time, but long-term results would be substantially better. I shall not dwell the advantages of Montesque's ideas for readers of the present work.

Separation of power in a multidimensional society, in turn, entails many view points. Of fundamental importance is the institutionalization of more or less equally powerful branches, both with respect to the scope of their activities, especially the separation of the ideological or religious function, and their territorial (central vs. regional power) and functional responsibilities. For the system as a whole, the last feature manifests itself in the separation of power between the bodies representing national unity, oneness, and national traditions, for instance, a constitutional monarch of the Anglo-Saxon variety, and other bodies that fulfill specific functions. The latter include, as part of the system of checks and balances, the legislative bodies, which lay down laws of fundamental long-range significance, constitution, for instance, and non-conflicting, more specific laws that are reviewed more often (parliamentary laws); the executive branch, which carries out these laws as required by the situation at hand; and the judicial branch, which sees that the laws, as well as the laws for changing the laws, are properly folowed, including preventive measures and injunctions. Interesting in this connection would the role of the fourth branch, empowered to make changes in the constitution. I merely wish to note here that the legislative branch, subject to certain provisions, does have the power to make some changes - amendments to the constitution.

These deliberations on the separation of power parallel our analysis of the social structure of biological systems as well as the role of different sexes. It is quite plausible that the mechanism of change is predicated upon multi-sexual reproduction which reflects the separation of functions fulfilled by the direct participants of the process of innovation. In effect, assuming that the male sex embodies the executive power and female sex the legislative power, a third sex could fulfill the function of the judicial branch.

## 1.2. The Pluralistic Mechanism

Consider the development of a political system. It is indeterministic by its very nature, and it operates based on a pluralistic mechanism. The objects of this mechanism are the various programs of development for a country. The pluralistic mechanism itself is characterized by the following features: 1) elaboration by independent participants of a manifold of programs of undifferentiated value, 2) selection of one program, or a combination of programs, to govern the country for a given period of time; this selection is based on the established procedure for setting the priorities among the various programs in the manifold, 3) supervision over the implementation of the selected program, and 4) replacement of the selected program if it proves inadequate or if the conditions change.

In more general terms, the objects comprising a pluralistic mechanism are not limited to the actual programs of development, but consist of inputs/outputs of the system, including operators who implement the proposed programs.

The process of biological development, which is indeterministic in nature, follows a somewhat similar course. It is carried out by means of a pluralistic mechanism whose objects are the living creatures.

In the initial stage, the manifold of living creatures is formed. This stage supports the creation of new creatures, their selection (preventive selection prior to birth), and ensures continuity between different generations necessary in view of the discreteness of mortal creatures. In the second stage, the created manifold of creatures "lives and works" in a specific environment, where the creatures' capacities to develop are actually realized. At this stage, as well as at later and earlier stages, the methods of operation of the living creatures change (the diversity of these methods expands), the best suited methods are selected, and, finally the actual performance of the selected methods is supervised and, if necessary, the methods are replaced. The third stage reveals how well the selected set of living creatures copes (and thus survives) in a given environment. In the fourth stage, the manifold of creatures undergoes restructuring with the ratios among the various species changing depending on the environment. Naturally, there is feedback from all these stages.

In summary, each stage reflects its own dynamics, which unfolds within the framework of a pluralistic mechanism.

The biological world is peculiar in this respect because the vehicles for implementing all these stages are the individual living creatures. This means that the creation of the manifold of living creatures entails not only changes in the anatomy or physiology, which was the primary focus of the evolutionists, but also the methods of their interactions with the environment. By methods of interaction I mean the capacity to develop new programs of behavior, including a) various mechanisms of evaluation such as emotions, biological drives, pain; b) ability to perceive the environment as well as the organism's own abilities; and c) "algorithms" to carry out the above. An organism possesses various physiological structures that support the development of its behavior. At the same time, there exist mechanisms for changing these structures.

Living creatures are capable of observing/assimilating the results of the interaction between their program of behavior and a given environment and modifying it through the process of learning. The learning program is fixed in the physiological structures of living creatures. Higher organisms, and man in particular, possess physiological structures capable of changing the learning programs.

Finally, if some species proves ill-adapted to a given environment, it might barely survive, yielding more space to other creatures.

From the process-oriented point of view discussed above, the volutionary process can be divided into the following stages: *creation* of the manifold of living creatures; *performance* - interaction of these creatures within a given environment; *revision* - evaluation of the creatures' performances in terms of their adaptation or changes in the manifold and finally; *replacement* - changes in the structures of the living creatures made possible by the expanding manifold and different external conditions (environment).

Each stage of the pluralistic mechanism incorporates change, selection, and heredity. Moreover, change encompasses not only the creatures themselves, but also the methods of selection and heredity, all implemented through the individual creatures. In our subsequent discussion, I shall use this broader notion of change. To the best of my knowledge, the currently prevailing view in biology states that change pertains only to the organisms themselves and that the mechanism of change is fixed, just as the methods of selection and heredity.

## 2. MICROLEVEL

### 2.1. Typology of Deviants

In the social realm, an individual attempting to change the established norms is labeled a *deviant*. I call a deviant who has a positive impact on society an *innovator,* and consider a deviant who is detrimental to society, as abnormal or sick.

There are two kinds of sick deviants - pathological and nonpathological. What distinguishes a pathological individual from a nonpathological one is the former's inability, given the resources available, to maintain normal functions and handle adverse situations.[4] Therefore, sickness should not be equated with pathology, since the latter implies one's inability to deal with one's ills. For instance, a person with the flu experiences both an abnormality and a sickness, but the state is not pathological if the body can deal with the sickness on its own.

In medical terms, deviants can be classified as *healthy* or *sick*, on the one hand, and *pathological* or *non-pathological,* on the other.

Other criteria can also be used to distinguish among different types of deviants. One is an ideology espoused by the deviants for transforming society and the degree of radicalism embraced by this ideology in terms of its claim to power and wealth. In a sense, these characteristics reflect general systems categories, which makes them applicable to biological systems. At the same time, "social deviants" might possess certain features that other systems lack, for instance, the perception of one individual by another.

The following table gives the taxonomy of deviants based on the parameters outlined above.

**TABLE 2.1.** Typology of deviants.

| Usurpation | Ideology | | | |
|---|---|---|---|---|
| | Present | | Lacking | |
| | Positive (innovator) | Negative | Positive | Negative |
| Desired | Revolutionary | Terrorist | Victim of a system | Bandit |
| Not desired | Reformer | Revisionist | Liberal | Opportunist |

Deviants in the social realm can further be classified dichotomically into two groups - conditional (situation-specific) and unconditional. Conditional deviants can be both positive or negative, depending on the situation. For instance, a reformer is regarded as a positive figure, while a revisionist has negative connotations. Therefore, the same deviant can be both a reformer or a revisionist, depending on the circumstances. There are also unconditional deviants, deemed so no matter what the circumstances are. For instance, bandits are always judged to be bad. My concern here is primarily with the general case of conditional deviants.

## 2.2. The Role of Deviants and Their Interaction

Every developed pluralistic political system includes the entire spectrum of deviants espousing their own ideologies. Let us take radically-inclined deviants. It is not unreasonable to suppose that their programs include some valuable points which, under certain conditions, can actually improve the operation of the system that is founded upon other principles. I would say that, in general, the programs proposed by radicals are inadequate relative to the complexity of the problems they claim to solve. The danger of radicalism is the desire to impose its ideology upon others forcibly. The influence of radical programs depends on the radicals' capacities for self-control, as well as existing circumstances; under stressful social conditions such as wars, economic crises, etc. society is more prone to extremism. The radical element does not pose a grave danger, even in time of crisis, if it is capable of exercising some self-restraint and the political culture of the populace is sufficient to prevent radicals from usurping power. Now, if one or both of these conditions is lacking, especially during stressful times, the country can easily fall prey to these radicals who include anarchists, communists, fascists, and similar groups. During the devastating crisis of 1929-1933, communists in the United States could not gain power. One reason was that they exercised some self-control. For instance, in keeping with American tradition, they put their own personal interests first – they did not strictly follow the *party line*, and they were not totally obedient to party discipline, as was characteristic of the Russian communists. [1] Moreover, the political culture of the American people kept them from succumbing to radical views. At the same time, certain ideas advocated by socialists and communists were assimilated by American democratic institutions. These included all kinds of government-run social safety-net programs to protect the *down and out* – progressive taxes, minimum wages, unemployment compensation, social security, etc.

Meanwhile, in times of crises, communist (revolutionaries) were able to seize power in Russia, as did fascists in Germany, Italy, and Spain. The primitive "genetic program" proclaimed by the radicals penetrated deeply into the fabric of the social organism. It distorted the structure of the country and its mechanism of government, and it damaged people to such an extent as to lead the country, within an historically brief time period, to virtual exhaustion with all the ensuing consequences.

Let us now take a look at the typology of people from the standpoint of their attitudes toward other members of society.

We can distinguish three groups of people: egotists, individualists, and collectivists. An *egotist* is distinguished by the fact that he pursues his own interests while disregarding the interests of others. A *collectivist* is one who surrenders his own interest to the interests of the whole. An *individualist* is one who puts his own interests at the forefront, but takes into account the interests of others. In this context, an individualist represents a *microcosm* because he embodies his own interests as well as those of others. Democratic societies seek individualists, while totalitarian regimes hail collectivists. Frequently, a totalitarian society gains a large number of collectivists, but as they become disappointed, they turn into hard-core egotists.

This typology of individuals suggests a similar classification for cells. By analogy, an organism has conformist cells, i.e., cells which reproduce cells like themselves, as well as non-conformist cells. The latter category includes *innovators* - cells that are conducive to the development of an organism. Cells-innovators could be regarded as normal, since the change they undergo is in harmony with the changes they induce in other cells, including, perhaps, germ cells.

Deviant cells can be abnormal, posing a major threat to an organism. Abnormal cells, on the other hand, can be non-pathological, that is. capable of self-rectification; a pathological case refers to cells that are unable to revert to their normal state and, at the same time, cause harm to the organism.

Innovator cells can affect the genome of other cells (see in the following chapters), thus transferring their own positive "ideology of development" to other cells. There are also abnormal radical cells which, instead of helping an organism, attempt to impose their doctrines in a simplistic brute force fashion. Radicals who try to impose their destructive genetic programs upon other cells become *terrorists*. There are also *bandit* cells. Both use force, but bandits possess no "ideology" and do not affect the genome of their victims. Bandits can act like murderers out of control, perhaps damaging other cells' membrane, or like murderer-burglars who kill selectively in pursuit of personal gain, perhaps taking the nutrients from the victim-cells.

In principle, radical cells might well carry valuable genetic information, but the only way this can be beneficial is if the organism's immune system, analogous to political culture of a nation, curbs these cells, thus preventing them from usurping control over the organism.

My natural philosophical approach runs contrary to common wisdom, which holds that each individual cell must serve the interests of the whole, sacrificing itself, if necessary.

In my philosophical scheme, there is a variety of cells. Some cells behave like *individualists*. A cell-individualist pursues its own interests and seeks its own development. Cell specialization benefits this cell because its chances for development improve under this system of mutual exchange. At the same time, each cell must take the "interests" of other cells into account. The genome of every cells reflects its individualism: it contains information about the entire organism, the cell itself, and the program of development of this cell in its respective environment. There are perhaps *altruist* cells, willing sacrifice themselves "voluntarily" to save an organism (*apoptosis*). There are also *selfish* cells which subjugate the development of an organism to their own narrow interests, even if their behavior eventually leads to their own demise.

I also want to note that cancer represents a systemic disease, an expression of a pathology of the propensity of living creatures for change. It is a manifestation of selfish, turned pathological, radical cells, of terrorists or bandit variety, aggravated by the fact that the organism's immune system fails to bring these cells under control and, at the same time, save innovation-carrying cells from the same fate.[2] It is by no means indisputable that cancer cells, as defined above, are completely unmanageable or that their growth necessarily unstructured. Their destructive behavior and growth could be tamed and structured and their destructive program rectified by the immune system.

My speculations suggest a number of hypotheses pertinent to the way we view cancer. Here, I shall limit my argument to one illustrative hypothesis .

Essentially, my hypothesis states that the behavior of an innovator cell is almost identical to that of a cancer cell, especially at the early stages. The confusion between these cells could be extremely injurious to the evolutionary process because innovative cells categorized as cancerous are subject to destruction. The medical profession today employs all means at its disposal to destroy cells deemed oncogenic or turn them into what is conventionally regarded as normal cells, i.e., stable cells which reproduce a given organism.

It is particularly dangerous to destroy innovator cells under early diagnosis of cancer, since the behavior of a cancer cell at an early stage may be quite similar to that of an innovator cell.

I believe that innovator cells that may turn into terrorist cells should to be treated differently. Assuming that the future course of development of such cells is uncertain, meaning that there is no guarantee that the "society" - the organism's inner environment - will be able to cope with the innovative cells in case they turn out to be terrorists, it might make sense to isolate such cells "from society" rather than destroy them. The latter task should be carried out by the immune system. Its role is to promote as much as possible an organism's biological potential for development by preserving the necessary diversity of cells, rather than destroying the harmful cells, or other harmful ingredients. Over the course of time, once it becomes clear that the body can control the terrorist cells, these cells could be allowed to function normally.

I realize full well that at this point in time, when we are unable to stop the onslaught of cancerous cells, my remarks sound rather out of place: when the organism's survival is on the line, thinking about its development seems frivolous (as the saying goes, "Thank God just to be alive"). But, perhaps, it pays to look ahead, and not sever the link between the development of an organism and the development of a species, but to bridge the mechanism of change within an organism with the general process of evolution!

## 3. THE LESSONS FROM THE SOCIAL MECHANISMS OF CHANGE FOR THEIR BIOLOGICAL COUNTERPARTS

What specific devices relevant to biological evolution and biological mechanisms of change are suggested by our analysis of socio-political systems?

The following ideas come to mind:

1) Separation of power in the political realm ensures separation of functions needed to govern a society with subsequent integration of these functions via a horizontal mechanism of checks and balances.

2) A pluralistic mechanism permits a society to combine strategy, concerned primarily with expanding the manifold of objects and methods of their interaction, with tactics concerned primarily with selecting from this manifold the set of alternatives that best addresses the current goals and with supervising the implementation of the selected program.

3) The vehicle of change at the microlevel is the diversity of deviants, ranging from extreme radicals to moderate reformers; all of these groups ought to be preserved. The ways of preserving this manifold are

contingent upon the circumstances peculiar to each country, particularly, the political culture of the nation.

The above considerations on the pertinence and instructiveness of socio-political ideas for the analysis of biological systems suggest the following hypotheses regarding the latter:

Hypothesis number one. It is quite plausible that the mechanism of change is rooted in multi-sexual reproduction that reflects the separation of functions fulfilled by different participants of the process of innovation.

Hypothesis number two. Evolutionary development calls for the preservation of the manifold of deviant cells, ranging from radicals to moderate imitators.

Hypothesis number three. Cancer can be viewed as a pathology of a normal somatic mechanism of change which involves radical deviants poorly controlled by the immune system. Since cells-innovators and cells-destroyers may appear to be very similar, especially at the early stages of the process of change, it is important, when diagnosing cancer, to make the distinction between these two types of cells. Moreover, under certain conditions cell-destroyers should be sequestered rather than eliminated.

## NOTES TO CHAPTER 2

[1]. Interesting in this respect is the American movie Reds. The hero of the movie is a real-life character, an American communist, John Reed. He was not just an observer of the Bolshevik Revolution in Russian, but, to a certain extent a participant. There is a scene in the movie which depicts a conversation between Reed and one of the leaders of the revolution, Gregory Zinoviev. The discussion centered around Reed's insistent request to go back to America to share Christmas with his family. Zinoviev thought Reed's presence was crucial for the revolutionary meetings held in Azerbaijan.

I am grateful to my American friend, Norman Gross, for clarifying the differences between Russian and American Communists.

[2]. Paraphrasing the hypothesis advanced by Galina Filonenko regarding the analogy between cancer cells and abnormal individuals, slowly progressing cancer is akin to pathological egotist cells of the evolutionary type. Terrorist cells are associated with cancerous cells which partake of the "normal" course of cancer; bandit type cancer cells partake of rapidly progressing cancer, sarcoma.

*PART TWO*:

## *Evolutionary Mechanisms of Change: Normal Case*

The set of biological objects, together with their mode of interactions, make up the biological system. Just like many other systems, biological systems are dynamic in the sense that they change over time. Considering the stability of an organism as far as its adaptability is to the environment, including its ability to "repair" itself, an organism can easily turn into a very conservative system incapable of change. However, organisms manage to overcome this conservatism, which is really a challenging feat. Ernst Mayr [37] has noted that

> "The real problem of speciation is not how differences are produced but rather what enables populations to escape from cohesion of the gene complex and establish their independent identity. No one will comprehend how formidable this problem is who does not understand the power of the cohesive forces in a coadapted gene pool." (p.297)

Let us examine the general characteristics of the mechanisms of biological change from the systems multi-dimensional perspective, i.e., function, structure, process, operator, genesis.

# CHAPTER 3

# THE MECHANISM OF BIOLOGICAL CHANGE – GENERAL CHARACTERISTICS

## 1. CATEGORIES OF BIOLOGICAL DYNAMICS

In keeping with the spirit of the present book, I shall distinguish the following categories of biological dynamics: its sources, its course, and the methods of representation which correspond to the various mechanisms of implementation.

### 1.1. Activeness and the Course of Development

Biological objects are *active*, meaning that they possess internal mechanisms which allow them to shape the environment as well change their own performance, depending on the signals received from the environment.

Under ideological monism instituted by those in power, the conflict between the active and the passive approaches to the behavior of biological systems may assume rather violent form.

In the Soviet Union, at the end of the 1940s, Pavlov's theory of animal behavior, with its emphasis on the passive mode of animal, as well as human, response through the development of reflexes to external stimuli, became the ruling ideological dogma in biology. The proponents of the opposing theory, which endowed animals and especially human beings with more capacity to control one's behavior,[38] experienced severe hardship. Only after Stalin's death were Pavlov's teachings dethroned as the only true theory and relegated to one of many important approaches to the study of animal (human) behavior.

Active systems, picked for methodological purposes as those determining biological dynamics, can be represented as having *random* or *directed* development. In the present context, the term random does not preclude the usefulness of local actions as long as these actions are not ordered.

**1.2. Survival, Viability, Growth, and Development**

The course of development of any system is defined by such categories as *survival*, *viability*, *growth*, and *development*. Survival connotes the preservation of a given organism *per se*. Viability is the preservation of a given organism via reproduction.[1] Growth implies a quantitative rate of change different from just one (one represents survival or viability); development involves qualitative shifts where each step is associated with a qualitatively new phenomenon requiring new methods of operation. By analogy with physics, compounds undergo phase transitions to assume qualitatively distinct states. The category of development as it pertains to biology must be qualified to distinguish development in the sense of "construction" (development of a biological entity based on a given program*)* from development which incorporates changes in the program itself. The first case is commonly known as *embryonic* development. The second type of development shall be called *innovative development*.

Along with Russell Ackoff and Jamshid Gharajedaghi,[39] I am a proponent of the primacy of development, with growth, viability, and survival allotted a subordinate role required to support development. Within this conceptual framework, the global formulation of the dynamics of biological systems becomes crucial. The view that the key objective of living creatures is survival is widespread. Survival, being a prerequisite for all else, is essential. However, if survival is deemed the ultimate goal, *local* needs tend to dominate, since survival today overshadows all other concerns. In principle, viability, growth and development can be viewed in terms of long-term survival, but it makes our analysis very cumbersome.[2]

Another factor affecting the assignment of priorities among the many alternative objectives of system's dynamics is the decision-maker's mind-set.

Perhaps an example will clarify my point. Some years ago I happened to meet some leading executives from the Clark Corporation. At the time, the firm was facing a crisis, it was, in fact, on the verge of bankruptcy. Under the circumstances, survival would seem the most pressing concern of the firm's top executives who were focusing the elimination of all unnecessary current expenditures. However, the mind-set of the firm's leader, James Rinehart, a man who has successfully combined academic training and practical thinking, was geared toward development; survival was viewed as a precondition. The corporate slogan became: *"Cut expenditures; don't cut your future."* As a result, cutbacks initiated by the firm were carefully screened so as not to hurt the firm's future. Many corporate leaders recognize this, but it takes a special gift to implement the idea of development and make survival and growth subsidiary, especially when the firm is threatened with bankruptcy.

## 1.3. Directedness and Goal

The terms *directedness* and *goal* are not identical. If all we were concerned with was the *course* of the system's dynamics, it would be sufficient, by definition, that there exist states sufficiently remote from the initial state. In Greek *tele* means 'remote'. Thus, an exclusive emphasis upon the direction of development could be termed a *telelogical* approach. Once a *telological* component is incorporated into the above scheme, we have a goal, *telos* in Greek, representing a terminal state toward which the system gravitates. Here, the category of a goal may appear implicitly, i.e., as a manifestation of the system's trend of development, or it can be formulated explicitly as one option within the set of available alternatives.

The two terms introduced by Ernst Mayr[40] in relation to biological systems - *teleonomic* and *teleomatic* processes - reflect these two methods of setting goals. In general, the term *teleological* is reserved for situations where directedness also incorporates the component of a goal.

While the study of biological systems claims to based upon causal rather than telological approach, it is implicit that the analysis is largely teleological, in the sense that *directedness* is postulated as the *struggle for survival*. It is this tenet which equates the performance of any biological entity, e.g., cell, organism, species, with its *eternal existence* that is implicitly assumed to direct the energy of the living system.

## 1.4. Representation of a System

Let us consider the various ways in which the course of biological dynamics can be represented.

The course of biological dynamics can be inferred phenomenologically via the *extrapolation* of past trends, in which case we search for the *laws* of development of a biological system represented as a *black box*. The implicit goal is to relate inputs and outputs, ignoring the actual mechanism of transformation formed by the established rules of interaction, e.g., physical, chemical reactions; technologies, among the objects comprising the system. It is further assumed that the link between inputs and outputs is represented in the most compact form, by a formula, meaning that the need to *select* subsequent course of development from *alternatives* at each step of the procedure does not arise.

The course of development can also be deduced based on non-random mechanisms of operation of biological systems, subject to some basic principles of interaction of biological objects comprising the system.

There are at least two ways to represent these principles of interaction. One representation of a biological system is based on the basic principles of interaction of its constituent elements that, given the initial state of the system, ultimately attain equilibrium via forces, attractors and repulsors, that define the system's dynamics. The second approach is predicated upon the assumption that system's dynamics obeys certain criteria of optimality that reflects the totality of forces acting upon the system, subject to constraints on the initial state and the rules of interaction. The first approach corresponds with models of equilibrium; the second one yields an optimality-driven representation of a system.

It is important not to confuse the method of representation of a biological system with its mechanism of operation. The latter falls into at least two categories: *horizontal* mechanisms based on the interaction of equal biological elements, and *vertical* ones predicated upon a governing body. Generally speaking, each mechanism of operation can be combined with each of the aforementioned modes of representation.

I would like to note in this connection that certain contrived artifices, when used *heuristically* may be conducive as well as obstructive to the development of a given field. The concept of God who rules the universe not only impeded the progress of science, but facilitated the development of certain great ideas, such as the extremal principle in mechanics characterized by treating system's dynamics as a search for a optimal trajectory.

As far as biological systems are concernd, the proponents of the scientific approach often attempt to create models of dynamic ecological equilibrium[41], while the adherents of the Divine origins of the universe, *creationists*, devise models with an explicit and well-defined criteria of perfection. Biological models complying with the rigid requirements of hard science may also incorporate the principle of optimality. An example are models of blood circulation.[42] Both approaches represent methodological devices for representing the system, and they should not be confused with the actual mechanism of operation built into the system.

As the history of physics and economics has shown, the confusion between the representation of a system and its mode of operation can wreak great havoc.

The proponents of the equilibrium model in physics thought they were describing a world of cause and effect with no God; the advocates of the extremal principle insisted on God's creating a perfect world that is governing by a certain criterion of optimality. Considering the time when the church in Europe was not separate from the state, this kind of ideological interpretation of the universe was perceived by scientists-atheists as not so innocuous. It took a long time before the affinity between the two concepts was understood mathematically. Then the conflict subsided. As a result of the separation of church and state, the discussion eventually shifted to the realm of science. Its ideology-free agenda was concerned with the various methods of representing the physical world.

In economics, the equilibrium representation was seen as a manifestation of spontaneous, chaotic, market-type development driven by many participants interacting with each other via the price mechanism. Optimization models were associated with a centrally planned economy governed by a central body which allocates economic resources according to the selected criterion. I do not think it is necessary to dwell on the damage caused to economic science by this kind of confusion, especially in former communist countries with a plan-based system rooted in the Marxist economic doctrine.[43]

It should be noted that even to this day some professors in Catholic universities maintain a keen interest in the extremal principle in mechanics. In fact, the Catholic church has expressed enthusiasm toward the active modes of representation of economic systems. The Papal Academy of Sciences published a two-volume set compiled from works presented at a scientific conference on the problems of economic planning.[44] The conference was held from December 7–13, 1963. It was attended by most economists from around the world who were involved in the problems of state control of the economy. The Pope himself appeared before the gathering.

Let us not, however, ignore the history of science and reject creationism outright. If we choose to interpret it non-literally, it could hint at some interesting new ideas.

The next problem to consider is the basic mechanism of the implementation of biological development.

## 2. FUNCTIONAL ASPECT: WHAT IS THE COURSE OF BIOLOGICAL CHANGE?

A telelogical framework seems to be an appropriate methodological tool for a related question: "What is a plausible course of development of a biological system?"

### 1.1."Apart From Trying to Survive, Every Species Strives to Create A Species More Perfect Than Itself"

We can construe a scheme to guide the development of the biological world that manifests itself in the following principle: *"Apart from trying to survive, every species strives to create a species more perfect than itself."*

This principle resembles the design approach to biological systems, and it has been discussed by a number of scholars in the field.[45] The approach is not all that absurd and, apart from its methodological significance, as a metaphor, it is helpful from the ontological perspective. In the light of recent discoveries in molecular biology, we now recognize that embryonic development is based on a program very similar to those governing the design of complex objects. This program incorporates the idea of a hierarchical procedure of constructing an organism. It was discovered that many different creatures, from *Drosophile* to man, possess the so called *Hox genes*[46] whose number fluctuates between 8 and 36; distant related genes have been found in plants, fungi, and molds. Hox genes are active at the early stages of embryonic development. They determine an organism's basic structure by telling the various cells their destination: where the head is, the chest, etc. Hox genes function by producing proteins called transcription factors. These proteins clamp onto the chromosomes and trigger, in a wave-like fashion, the action of subordinate genes. These genes operate in a manner of simple information signals, that is they produce a strong biochemical reaction induced by a rather weak impulse. The subsequent development of individual organs is another problem with its own long history.

Let us return to this paradigm that, "Apart from trying to survive, every species strives to create a species more perfect than itself."

Assuming that evolution is progressive, one observes that living creatures have undergone certain improvement. However, it seems impossible for a species to be more perfect that another species if all relevant parameters are taken into account. Some obsolete species which have given birth to a new species, might still not perish. Considering the fact that the environment is far from stable preserving the new species as well as the old provides considerably more space for living creatures to develop. Naturally, the ratio between the old and the new species comprising the manifold of creatures changes over time; most species disappear altogether.

Once we accept the above paradigm, the struggle for survival - the guiding hand of evolution - as well as growth ( in the sense of change, with positive or negative signs, in the size of the population) become necessary conditions for the process of development.

Man might very well not be the crowning or the end point of development and just as Man evolved from an ape-like primate, so a new species springing from Man may appear in the future.

These are the thoughts articulated by Nietzsche, through Zarathustra in his speech to the people:

> "I teach you the overman. Man is something that shall be overcome. What have you done to overcome him?
> All beings so far have created something beyond themselves; and do you want to be the ebb of this great flood and even go back to the beasts rather than overcome man? What is the ape to man? A laughingstock or a painful embarrassment. And man shall be just that for the overman: a laughingstock or a painful embarrassment. You have made your way from worm to man, and much in you is still worm. Once you were apes, and even now, too, man is more ape than any ape.
> Whoever is the wisest among you is also a mere conflict and cross between plant and ghost. But do I bid you become ghosts or plants?
> Behold, I teach you the overman. The overman is the meaning of the earth. Let you will say: the overman shall be the meaning of the earth!"[47]

## 1.2. How Can one Create an "Overman"?

There are at least two ways to proceed: one is by continuing the course of biological evolution, and the second is by artificial means.

Let us consider the first route. Its mere feasibility is important for my purposes: allowing for future biological changes justifies my hypothesis of an active mechanism of change instilled in human beings.

Philosophy aside, let us see what biologists and non-biologists have to say about the future evolution of man.

Scholars most firmly entrenched in the current scientific mindset, which limits their horizon to about 50 years, are rather circumspect in predicting the future of man. They do not expect any major changes in the human species in spite of drastic changes in the environment. Some scholars express more concern with such factors as the food supply to feed the exploding global population. Many prominent scholars came to Dublin in 1993 to discuss these issues at the conference entitled "What is Life?". A brief synopsis of the conference was presented in O'Neill, et al. [48]

Free of the shackles of the current state of science, I am going to let my imagination run wild and look at the distant future of the human race.

Practically all scientists believe that man will continue to evolve perhaps eventually creating a new race. A famous French astronomer Camille Flammarion (1842-1925), had the following thoughts on the subject:

> "A new race, intellectually more developed, shall take our place on Earth, and who knows if you, my pensive and dreamy readers, and I are not destined to meet in an office of some scientist of the 276-th century as pallid and magnificent skeletons with name tags on our brows... We'll be looked upon as rather curious specimens of some long extinct race, fairly crude and vicious but possessing certain rudiments of culture and civilization and exhibiting a mildly pronounced proclivity for science."[49]

While most scientists acknowledge that there is still room for change, some suggest that human evolution has come to an end, at least as far as morphology is concerned. This claim is supposedly corroborated by the fact that man has become an absolute master of the animal kingdom and has therefore overcome the need to struggle for survival. In other words, while the possibility of future change is there, the proponents of the above theory would not classify these changes as evolutionary, in the sense that these changes are not dictated by an unyielding struggle for survival and subsequent selection.[50, 51]

I believe this kind of rejection of enduring evolution is due to a rather superficial notion of evolution reduced to the struggle between man and animals. It ignores the fact that man still struggles with a myriad of microorganisms and the victor is by no means certain, as well as with the inorganic world which continues to wreak havoc in the form of earthquakes, floods, and other tribulations. There is really no end to the human struggle for survival as long as man is what he is. The external aspect of survival is really just one side of the coin. Man also strives to improve his performance in order to grow (multiply) or otherwise develop within the limited resources available to him.[3]

The great majority of scholars believe that man will continue to evolve. This view is really based on the extrapolation of phenomenologically observed changes in human beings over time – from Pithecanthropus - Neanderthal - Cromagnon - to modern man. Extrapolation is also applied to individual body parts/organs showing that they may disappear[52, 53] or undergo structural changes, at least in quantitative terms, over time. Interesting in this respect are the observations made by the paleontologist S. Williston, whose findings were subsequently generalized by W. Gregory into what is known as "Williston's Law." Essentially the assertion is the following: as the organism evolves, polyisomerism (large number of elements comprising some part of the body) is replaced by anisomerism, or a small number of elements. Consider the monkey's tail. Man has retained four-five semi-reduced vertebrae of the coccyx and the number is sometimes as low as three.[54](p.78)

Within the phenomenological framework, the study of the various deviations from the norm found in modern man reveals interesting possibilities for predicting man's future. Pathologists have accumulated a much data on the variability of human body parts. It turns out that "under close inspection every single organ reveals some kind of deviation from the so called normal structure."[55] (p.77)

The major problem of the phenomenological approach is the time frame within which changes take place. Changes in the population as a whole are extremely slow, but frequently they manifest themselves in the anomalies found in individual beings:

"...the human organism is not an immutable or finished product. It is but a stage in the continuing evolution of man. Changes are slow. They take place over long periods of time and therefore do not yield to direct observation. It is only through anomalies and deformities - the key milestones - that evolutionary changes reveal themselves to us. At the present time, the course of human evolution is unclear. It is our job to provide a detailed description of all the anomalies and deformities that we find. Armed with this raw data future scholars will be able to know exactly, rather than just guess, the course followed by the philogenetic changes in our bodies."[56]

This methodological device (the role of abnormalities) will be helpful when I examine the mechanism of change itself.

The phenomenological approach to human evolution branches into many different mechanisms of change which I shall touch upon in the next chapters.

A new Man who will appear in the course of biological evolution might, at the very least, possess a much greater capacity to fulfill those functions that are intrinsic to man. One way to create such beings is by developing the already existing organs, provided it does not interfere with other functions. However, this course of human evolution is slow; in fact, it is significantly slower than in other primates and mammals.[4]

More radical ways of human evolution - ones not necessarily rooted in biology - are also plausible. Improvements might be achieved by combining artificial and natural organs. Biological evolution characteristically replaced or strengthened certain organs with external means; for instance, a natural shell replaced by some artificial cover; the use of sticks, stones and other artifices, instead of strengthening the body parts themselves.

In this respect Man has come a long way compared to other animals. Man has created rather powerful devices to improve his extremities (arms, legs) and to be able to do things for which he lacks specialized parts altogether, flying, for instance. Man has also begun to create artificial devices capable of improving or even replacing certain internal organs, e.g., kidneys, heart. It seems that this process of substitution is boundless, and, eventually, Man, made up entirely of artificial internal and external parts, could be created. The new type of artificial Man really represents a new species, since he will reproduce based on principles completely unlike those underlying human reproduction. One name for this new species is *Kiberhomo*. It is a combination of cybernetic technology and human structure.[5]

Thus, a new species superior to Man could be created on the substratum of Man. The creation of a new species on the substratum of another species with previous evolutionary experience taken into account is characteristic of biological evolution. However, it might be rather painful and slow in terms of evolution to create a new "improved" species based on the same principles as those underlying human machinery.

Perhaps the most effective way to create a new species is to invent new principles completely unlike those governing human development. Such a species could be created by Man outside of himself, as an artificial system. Man-made technology based on the new principles could eventually turn into a self-developing autonomous system having considerably greater creative powers.

Two questions arise in this respect: 1) Is an artificial system capable of formulating its own goals?, and 2) Can Man assign goals and constraints to this system so that the side effects of its operation will not cause him too much suffering?

There is no definite answer to these questions. People with the so called Western system of values continue to develop artificial systems capable of being superior to Man. They assume that the point of irreversibility, in terms of the welfare of mankind, is still very remote. In this sense, the disciples of Western civilization, no matter what local benevolent goals they advocate, are in a global sense, following the teachings of Zarathustra: "What is great in man is that he is a bridge and not an end: what can be loved in man is that he is an overture and a going under." [57] (p.27)

These thoughts stimulated me to propose the following interpretation of the Old Testament: God is not an absolute, but rather a developing entity; Man was created by God in order to augment God's greatness. We can further surmise that Man can increase his own power and eventually create a force superior to himself and increase God's might at the same time. This way God's might grows faster than that of any of his creations (or the creations of his creations), preventing them from ever becoming greater than God. Here, it is time to stop, for these deliberations fall far outside the scope of the Old Testament.

As we examine the *structural* aspect of the mechanisms of change it is important to pinpoint the type of cells we are talking about, namely germ and/or somatic cells. Whether or not changes in the somatic cells are heritable is another sensitive issue.

Much of my discourse regarding the mechanisms of change pertains to both somatic and germ cells; in case certain ideas apply to one type of cell only it shall be so stipulated). The reader is warned to shy away from assuming that changes associated with somatic cells must necessarily pass on as hereditary information. I realize this temptation, for, I, myself, would not rule out that under certain conditions changes in somatic cells do affect germ cells.

## 3. STRUCTURAL ASPECT: THE STRUCTURE OF THE "GENERATIVE SYSTEM"

### 3.1. "Generative System"

Biological mechanisms of change are usually associated with genes. To analyze this fallacy, I need to revert to the definition of the vertical and horizontal mechanisms discussed above. These categories, presented in the context of economics, have general systems significance. Equally relevant to our discussion of the biological realm,[58] these categories will help devise an adequate representation of the general structure and dynamics of the process of change.

It can be reasonably surmised that, at the very beginnings of the evolution of life, nature operated with a rather limited number of primary biochemical compounds and (bio)chemical reactions. The main problem under the circumstances was to augment the set of chemical compounds and methods of their interaction. Here, the horizontal mechanisms seem to have predominated. They reflected direct interaction of various elements and biochemical compounds without the intervention of a specialized coordinating program. This perspective on horizontal mechanisms operating in biological systems is characteristic of the theories of self-organizing systems, such as the theory of autopoiesis.

Just a passing note based on general considerations on the development of the universe and the evolution of living creatures: it quite plausible that survival, growth, and even the development of living creatures, of which change is an integral part, could unfold via basic chemical reactions without such specialized informational background as RNA and, subsequently, DNA.

As living system became increasingly more complex (greater variety of biochemical compounds and methods of their interaction), the evolution of life was driven by new combinations of existing primary compounds as well as methods of their interactions. The huge number of possible combinations necessitated some kind of a systematic procedure of selection aimed at eliminating large parts of inherently ineffectual genetic combinations. This is where vertical mechanisms, initially, perhaps, operating, from the structural point of view, through the hormones [6] and subsequently through the genes, come in as programs governing the creation of new organisms.

The emergence of the informational substrate (*the genetic system*) marked a tremendous step in the evolution of the organic world. It made several things possible including the capacity to preserve "memory" of the past in a very compact form, to assimilate incoming information, and to simulate the situation by means of built in programs. An informational profile introduced anticipatory techniques to regulate the behavior of compounds that form living creatures and to send signals that organize the behavior of living creatures.

It should be noted that the concept of a genetic system is more general than that of the "*genome.*"

As a rule, the process of change is associated with the genome as one complete haploid set of chromosomes, which carries a program of the creation of an organism. However the set of genes involved in the development of an organism is not necessarily limited to the genome. In fact, active genes may be housed outside of the cell's nucleus - in organelles, including mitochondria, or they may scattered in the cell's cytoplasm in the form of viruses [7] and similar structures.[8] The term "*genetic system*" encompasses the entire set of genes be they in the nucleus or the cytoplasm. Our discussion of biological change will be limited primarily to the genome. I should note that there are two kinds of changes in the genome, which I call operational and structural. The operational case implies that certain genes are suppressed or activated with the structure of the genome remaining intact; under structural changes the structure of the genome itself is altered.

The above model of the genetic system reveals that in spite of "verticality" it is not completely centralized. There are independent genes located in parts of the cell other than the genome.

The notion of vertical genetic mechanisms in the genome predominating in the process of change does not rule out the existence of other mechanisms.[9] As is the case with any innovation, the informational profile does not nullify direct interaction of chemical compounds but merely limits its scope.[10] Horizontal mechanisms are manifest in biological innovations being driven not only by changes in the genes but independently by changes in the cytoplasm.

Theories of the hereditary function of cytoplasm, along with chromosomes, have a history that is more than a century-old and is well known in the literature. Although chromosomal heredity has maintained its primacy and many attempts to explain heredity through cytoplasm have been rejected, the general concept of cytoplasmic heredity (non-Mendelian inheritance), as Ernst Mayr mentioned[59] (pp. 786-790), is still very fertile.

As early as 1950, T.Sonneborn published a paper entitled "Partner of the Genes." He elaborated upon his ideas in an article published much later in 1979.[60,61] S.Løvtrup analyzed non-Mendelian inheritance in the book Epigenetics.[62] Other interesting findings in the vein of non-Mendelian inheritance were presented by P.Sheppard; one such discovery revealed the impact of cytoplasm on the resemblance of cuckoos' eggs.[63]

The latest advances in biology impelled Brian Goodwin to state unequivocally that the totality of the organism and the environment take part in the creation of a new organism. "The position I am taking in biology could be called organocentric rather than genocentric."[64]

To sum up, the mechanisms of change incorporate both vertical and horizontal structures that include genes as well as other components. The totality of all these elements could be termed a "generative system". This totality is founded upon horizontal mechanisms, but it incorporates powerful vertical ones as well.

## 3.2. The Hierarchical Level at Which Changes Take Place

Another aspect of the mechanism of change is the level of the hierarchy, i.e., the degree of aggregation, at which changes take place whether they are induced by internal or external sources, or are random or ordered.

We can distinguish at least three levels of the hierarchy: genes, chromosomes, and cells. Each of these structures represents an aggregate, meaning that changes in each respective structure do not necessarily affect any other structures. In present work, I am more concerned with the genes, but I shall touch upon other levels as well.

The source of change within an aggregate can be external or internal. Internal sources of change are associated with a mechanism of change that is self-induced, i.e., triggered by internal forces; external sources, by definition, affect a given aggregate from the outside. Each source is sufficient to induce change in an organism. My primary interest is with the internal sources of change, but I shall briefly discuss the external case as well as the combination of the two.

The process of change within an aggregate can be random or ordered (of course, this dichotomy ignores the degree of order). My focus here is primarily on the ordered sources of change.

The boxes in the table below correspond to different combinations among the levels of the hierarchy, the sources of change, and the binary degree of order of the process of change.

TABLE 3.1. Sources of change and degree of order.

| Structure | Process of change | The source of change ||
|---|---|---|---|
| | | External | Internal |
| Genes | Random | Chemicals, radiation, break downs in the cell | Damage |
| | Ordered | Viruses, regular fluctuations in radiation | Program-changing program |
| Chromosomes | Random | | |
| | Ordered | | Transposons |
| Cells | Random | | |
| | Ordered | | Crossing |
| Tissue | Random | Infection | Damage |
| | Ordered | | |

Some clarification is in order.

*The gene level.* This level actually comprises the primary subject matter of the book. At this point, I would like to note that viruses also count as structures that induce change at genome level.

These genetic strands merit a digression based almost entirely on the work by Konstantin Umansky.[65]

The term *virus* carries a negative connotation because of its association with various diseases. The leitmotif of Umansky's book is the positive role of the viruses, which manifests itself primarily in helping organisms to adapt. The diseases associated with viruses ought to be treated as a pathology of their positive functions in helping with adaptation.

> "...one feature shared by all respiratory virus infections is that they are seasonal and correspond to changes in the environment (Fall/Winter and Spring seasons). It is important to note that these outbursts are not "calendar specific" but correspond to the extremum points of the changing environment, i.e., time frames when adaptive reorganization is most urgent, especially so for respiratory organs. These observations lead us to conclude that certain respiratory viruses are factors that partake in the organisms' adaptive acclimatization." (p. 30)

For instance, the virus *Fignia* is a genetic factor that controls the sensitivity of certain types of *Drosophila* to carbon dioxide. When this virus penetrates the fly's genetic structure, it allows the fly to adapt its breathing habits to that particular environment. Characteristics thus acquired are passed on, establishing continuity so crucial for survival.

The feature that makes it possible for viruses to partake of the process of adaptation to changing conditions is their "omnipresence." It is rooted in the universal structure of the viruses, i.e., the ability to penetrate any living creature, including plants, animals, and man, and to migrate from one to another.

As a result "...viruses constitute the public genetic pool that can be used by any biological entity." (p. 12)

In other words, the diversity of viruses both inside and outside the organisms represents a gene pool; genes fusing with certain structures of the host cell lead to changes in the organism.[11]

*The chromosome level.* Changes at the chromosomal level are less explored than changes at the gene level. Here, I would like to quote from a work by Nikolay Vorontsov:

> "It is important to note that chromosomal rearrangements of the Robertsonian type (2Afi1M), and chromosomal inversions (i.e., rotation of an individual segment by 180º) do not alter gene composition; here, evolution does not effect the structure of genes and takes place without gene mutations. Consider the following analogy: gene mutation involves changes in hereditary constitution, which can be compared to changes in

MECHANISMS OF CHANGE: GENERAL CHARACTERISTICS 67

the content of a tape recording; chromosomal mutations, on the other hand, are like changes in the construction of the cassette carrying the tape. Chromosomal mutation changes the dispersal of hereditary information, because information flow ('gene flow') between tape recorders with differently constructed cassettes is bound to be limited."[66] (p.181)

More recent research on chromosomal mutations is linked to telomeres. This topic is discussed in greater detail in Chapter 6.
*The cell level*. See Chapter 6.
*The tissue level*. Change at the tissue level has multiple appearances. One of the more interesting forms of change is metaplasia. "The term metaplasia denotes a form of regeneration which ultimately produces a new tissue that is morphologically and functionally distinct from the original tissue."[67]

## 3.3. The Temporal Hierarchy of the Mechanisms of Change

We have looked at the structure of the genetic system from the *spatial* perspective. Mechanisms of change also have a *temporal* aspect, i.e., the time span of the change relative to the life span of a given organism. The temporal dimension can be roughly divided into three segments: current, mid-term, and long-term.

Current refers to shocks, sudden changes, short-term changes in temperature, etc. The common method of adaptation, or adjusting an organisms for homeostasis, to short-term changes is through flexibility incorporated into the various organs, i.e., by quantitatively varying the magnitudes of parameters governing the operation of the existing organs. These changes can also be induced by viruses which alter the DNA or the RNA of the cells.

Mid-term changes are accommodated in a more complex manner, possibly requiring restructuring of certain organs with corresponding changes, at least operational changes, in the performance of the genetic system. Structural changes in the genetic system are possible, although they might not become genetically fixed and would thus not pass on.

Finally, long-term changes in an organism can be tied to its development achieved through an active conquest of the environment and adaptation to some of the more consistent fluctuations. This kind of change is usually associated with major structural shifts in the genetic system, including the creation of new families, orders, classes, etc., with new genetic information being passed on to the progeny. Perhaps long term changes are implemented by means of an internal mechanisms of change at the level of genes and chromosomes.

The present work focuses on this last type of change. Current and mid-term types of changes ought not to be slighted, since their mechanisms of implementation might be similar in some respects to the mechanism of long-term change. For example, changes that take place through germ cell recombination can be important any time.

In generalizing what I have said regarding the temporal and the spatial hierarchy of the mechanisms of change, we have to look at the following structures:

Frst, the mechanism of change operating at each level of this spatial/temporal hierarchy, i.e., the interaction among the various structures of both the temporal and the spatial hierarchy;

Second, the mechanism linking the various levels of the spatial and temporal hierarchies, i.e., aggregation and disaggregation[12]; and,

Third, the mechanisms for changing all of the mechanisms described above.

## 4. PROCESSIONAL ASPECT: A POSSIBLE LINK BETWEEN THE MECHANISMS OF CHANGE AND SELECTION

The prevailing philosophy of the mechanisms of change postulates that all change is random, induced by external factors. Subsequently, forces, also external to the genetic structure, carry out the selection of the generated mutations.

Let the starting point of our analysis of the mechanisms of change be such elementary components as molecules and biochemical reactions which give rise to an organism. Under these assumptions, the entire problem of evolution is reduced to ordering the process of *selection* of living creatures from the possible combinations of these building blocks.

Kauffman writes: "We have come to think of selection as essentially the only source of order in the biological world. If "only" is an overstatement, then surely it is accurate to state that selection is viewed as the overwhelming source of order in the biological world."[68] (p.6)

By encompassing a wide array of modern discoveries in such areas as mathematics, physics, chemistry, and biology, Kauffman makes a strong case for the idea that evolutionary selection can be viewed as *self-*

*organizing.*[13] This innovative approach questions the undue emphasis, prevalent among biologists from the time of Darwin, on the random nature of evolutionary change produced by external factors.[14]

Taking Kauffman's framework as our starting point (the primary objects of a system are exogenous, i.e., determined by an outside world and the rules of their interaction are given), the problem of development does reduce to uncovering the laws of selection, or "algorithms" of selection based on certain criteria. Thus, selection determines which discrete objects (living organisms) shall exist in time and in space and which shall perish.

We can approach selection as a hierarchical process, meaning that it takes place at the innerlevel (among parts of a single organism) as well the interlevel (among different organisms). This hierarchy of selection forms a single chain, but the sequential nature of the process (selection at one level of the hierarchy preceding selection at other levels) suggests we use two terms - *change* and *selection* - to underscore the peculiarities of each stage. The term change will be reserved for the process of selection within the discrete entity. It is the first link of the selection process. Selection proper denotes interaction between the discrete entities and the environment.

The greatest progress in biology has been in the theory of selection, with heredity a close second. Mechanisms of change within the organism are relatively less explored.

A brief remark on the relationship between change and selection. The emphasis is placed upon constituent elements and random processes of transformation, while the actual selection of genetic combinations that generate a given organism is secondary. This approach would make sense if the number of possible combinations were relatively small, i.e., if all the combinations could be constructed and tested through selection. However, if the number of combinations is large, this kind of mutation mechanism is rather resource and time consuming. The other problem encountered in connection with the above approach is the creation of intermediate combinations in the absence of an end-goal – merely a vague course of development. I have touched upon this situation in connection with the tunnel process.

I shall illustrate this speculative point regarding the role of change and selection with an example from the field of artificial intelligence as applied to music.

The structures of a number of music genres, such as fugues and certain dances, have been formally dissected to such an extent as to yield to computer generation. By randomly varying certain parameters defining the structure of the musical piece, a computer can create a multitude of musical mutations. All of these mutations comprise *feasible space*, meaning that they conform to the criteria defining a given genre. The problem is how to select from this a practically infinite number of mutations. There are two ways to carry out the selection process.

One way is to have a *law* that governs musical composition. The law implicitly incorporates both the composer's taste, and, indirectly, listeners' taste, and external constraints such as technical characteristics of the instruments, accessible decibel range, etc. However, it is very difficult to formulate such a law. For one thing, the audience's taste is subject to change, and forecasting shifts in taste is virtually impossible.

In the absence of such a law governing musical composition, we can employ a criterion of selection, so that there are established priorities of one type of musical mutation. Presently, neither the composer nor the listener is able to formalize these criteria for selecting musical works of choice. Naturally, these criteria are employed at the intuitive level of the composer/ listener.

Assume for the moment that the composer knew the criterion for selecting the best musical work in a given genre. By conducting an exhaustive search of all possible mutations within that genre, the composer would be able to choose the best one. One limitation of this approach is that it is too time-consuming, especially if the number of "mutations" is very large, for example, all variations distinguishable by man. In other words, the criterion, even if known, is insufficient to select among a huge number of variations. This situation is similar to solving optimization problem: one also needs an algorithm of selection which deletes branches that will undoubtedly miss the sought after solution. It is well known that the ordered procedures of shrinking the feasible space containing the solution can be combined with random techniques such as the Monte Carlo method. In fact, strictly optimal solutions are not always attainable, if only because of temporal or spatial constraints of existing algorithms. Herbert Simon coined the term *satisficing* to denote this kind of given the constraints solution. In any case, unlike the listener, the composer does possess an algorithm of selection. This algorithm is generally intuitive and indeterministic. Of course, knowledge of musical theory introduces some order to this process, with some parts of the musical work possibly produced in a deterministic manner.

Let us now return to biological systems. By analogy with music, imagine a mechanism of change which generates a certain number of new

MECHANISMS OF CHANGE: GENERAL CHARACTERISTICS    71

mutations within a given genetic structure. It is quite possible that the genetic structure has a hierarchy of levels marking the difference among the species, genus, family, class, phylum. Within each level there is room for considerable change and variation. Moreover, within this space, nature seeks satisficing (rather than optimum) mutations either through a law or some criterion combined with an algorithm. These search techniques may well incorporate random mutations similar to Monte Carlo. Of course, random variations thus produced should be distinguished from the random mutations resulting from damage to the genetic structure.

The following table illustrates in a systematic fashion the relationship between selection and change.

TABLE 3.2. Ordering methods within the genetic structure.

| Search method | Random element included | |
|---|---|---|
| | Yes | No |
| Random | 11 | - |
| Ordered according to: | 21 | 22 |
| Law | 211 | 221 |
| Algorithm | 212 | 222 |
| Exhaustive Search | 2121 | 2221 |

It is reasonable to assume that whatever the actual mechanism of implementation of the process of change (through somatic cells or germ cells) may be, the cell's genetic structure is *selective* in creating new forms.

## NOTES TO CHAPTER 3

[1]. "There are two common misunderstandings of evolution that should be mentioned explicitly. The first is that evolution promotes survival. This is only indirectly true. What evolution promotes is reproduction. Survival is obviously important to reproduction..." [69] (p.22)

[2] Compare the Ptolemean geocentric and Copernican heliocentric views of the solar system. Both theories have "prophetic powers" to predict the position of the sun and the planets, but, while Copernican system does so directly and elegantly, the Ptolemean framework is rather artificial and cumbersome.

[3] Similar views argued somewhat differently (for instance, ecological changes due to industrial production) are expressed in the article by C.Wills.[70] The article was written as a rebuttal to J.Jones, a British geneticist. The latter published an article in

the New York Times in the mid-1992s in which he claimed that human evolution has come to an end because of a significant decrease in evolutionary pressure.

[4]. However, this course of human evolution is slow; in fact, it is significantly slower than in other primates and mammals. The relative pace of evolution was pointed out as early as 1961 by a molecular evolutionist, Morris Goodman. The hypothesis was confirmed by recent research in this field reported in the conference commemorating Goodman's 70th birthday: "Molecular Ánthropology: Toward a New Evolutionary Paradigm", March 12-14, 1995, Wayne State University, School of Medicine, Detroit."
More on this in A.Gibbons. [71]

[5]. The term is not my invention. I read about it or heard it many years ago from a friend of mine in the Soviet Union, Leonid Grinman.

[6]. The relevance of this comes to the fore in the light of recent discoveries purporting that the development of organisms depends not only on the informational structures such as DNA or RNA, but also on the equally important hormones which probably preceded the informational substrata in terms of the evolutionary timetable.
One of the champions of this approach to the evolution of life is Fernando Nottebohm, a prominent American behavioral ecologist and neurobiologist, who has explored the behavioral aspects of animals in an ecological community.
Let me quote the following passage from an article written by Natalie Angier:
"In the era when genetics and the glories of DNA reign supreme, and most molecular biologists are fixated on discovering the genes for everything from senility to shyness, researchers of a more naturalist bent are suggesting a different tack.
Genes are only a part of the story of any animal's profile, they say, and other influences, like hormones, can contribute to, complicate and in some cases override the innate program inscribed in a creature's genes. And although researchers have long appreciated that a fetus's own steroid hormones, produced by its growing sex organs - testes in a male, ovaries in a female - will in turn help shape the growing animal's body and brain, only recently have they paid significant attention to hormonal contributions from the mother or, in the case of litters, the other siblings in the uterus."[72]

[7] Viruses in the cytoplasm are interesting for exploring the link between non-nuclei genes and the process of change. There is indirect evidence, based on the rather contradictory claims concerning the causes of viruses, which suggests that viruses possess powerful structures capable of influencing the process of change. "Three hypotheses have been advanced to explain the origins of viruses: 1) viruses come from primitive pre-cellular forms of life; 2) viruses are degenerated microorganisms; 3) viruses are by-products of cell components that have escaped from under the cell's control." [73] p. 62.

[8]. "Suppressor gene called DCC (for deleted in colon cancer) may be as far from the nucleus as it's possible to be: at the cell's outer membrane, where it could be involved in cell-cell adhesion."[74]
A similar phenomenon occurs in other types of suppressor genes. The development of genetic engineering has reaffirmed the fact that the genetic system might incorporate structures outside the genome. For example, Duchenne's muscular

dystrophy results from a damaged gene responsible for producing a certain protein. Once the damaged muscle cells were injected with DNA luciferase expression vector the protein was produced. Moreover, the "transplanted" gene did not need to be in the cell's genome; it could have been part of the nucleus or even cytoplasm.[75,76]

[9]. The following fact may serve to reinforce this statement. The crossover between a mare and an ass produces a mule, while the crossover between a stallion and an ass produces a hinny. These two different mutations belong to the same species that is, in fact, infertile (only reversible fertile). Taking into account that the male germ cell contributes only the chromosomes during fertilization of female germ cell, while the female germ cell features an entire array of hereditary structures, the decisive role of the chromosomes in defining the species comes to the fore; distinguishing features result from the structures in the female germ cell.

[10]. n this connection, I recall a conversation with my friend, a physiologist involved in the study of trophic ulcer. His research into the link between the nervous system and the humoral mechanism at tissue level is pertinent to the study of genetic programs and self-organization at the cell level. One argument in favor of one particular method of treating trophic ulcer was based on the assumption that the phase of development of a tissue is correlated with the engagement of the nervous system in regulating the tissue. Trophic ulcer represents degenerated tissue, so its incessant regulation by the nervous system may be inadequate and actually impede recovery. This approach suggested a new method of treatment: to arrest the nervous system using Novocain and induce natural self-organization within the tissue; meanwhile, administer artificial feeding to the tissue using various lotions. Once the process of self-regeneration reaches a certain level of organization the novocain blockade is suspended reactivating the control by the nervous system (the tissue reverts to its normal metabolic mode).

[11]. These considerations regarding the positive role of the viruses in the process of change apply to many different kinds of viruses, including retroviruses."Based on morphological, virological, biochemical and molecular biological data, it is proposed that the presence of endogenous retrovirus particles in the placental cytotrophoblasts of many mammals is indicative of some beneficial action provided by the virus in relation to cell fusion, syncytiotrophoblast formation and the creation of the placenta. Further, it is hypothesized that the germ line retroviral infection of some primitive mammal-like species resulted in the evolution of the placental mammals."[77]

[12]. In a large scale system such as economics, changes occurring at different <u>levels of aggregation</u> is a key problem. Macroeconomic changes, i.e., changes in highly aggregated sets of indicators, must be represented differently at lower levels where the actual integration of these aggregates takes place. A lot of ingenuity was required to develop constructive aggregation/disaggregation techniques.[78]

[13]. There is a great deal of literature on self-organizing systems. I am, however, surprised that Kauffman's above mentioned book fails to mention any literature on the subject which was pioneered by Heinz Von Foerster.[79] It was further elaborated and applied to biological systems in the works of Humberto Maturana and Francisco Varela,[80] Milan Zeleny,[81] and others.

The other source of ideas in this field is the theory of automata developed by Michail Tseitlin,[82] Victor Varshavsky and Dmitry Pospelov.[83]

It seems to me that whatever Kauffman's own opinion of the aforementioned works is, he should have made it known, since the thrust of his own work follows similar themes.

[14]. Kauffman also applied his concept to economic issues and attempted to link internal mechanisms of technological development, which belongs in the realm of engineering, with the selection of the actual technologies, which is subject of the economic science proper.

# CHAPTER 4

# TWO CLASSES OF EVOLUTIONARY MECHANISMS OF CHANGE

## 1. MECHANISMS OF SURVIVAL, VIABILITY, GROWTH AND DEVELOPMENT

One way to classify biological mechanisms is according to the objectives pursued whether they are aimed at sustaining an organism (survival), or at an organism's growth and development.

There exist two different mechanisms for sustaining an organism. One aims to keep the entire organism alive by repairing or replacing its malfunctioning parts. Such common biological phenomenon as repair, particularly in cells, and regeneration of certain body parts attest to the viability of this mechanism. While the repair methods are not the main tool in preserving the species, they are still of paramount importance. Assuming that the task of the living organisms at some stage of the evolution was limited to self-preservation (as opposed to quantitative growth or development), this class of repair mechanisms could have played a leading role. It is entirely possible that, looking at the evolution of the physical world as a whole, living creatures who lack the capacity to expand (in number) but have internal mechanisms of self-preservation, might turn out to be more durable.

The other mechanism of survival is predicated upon the mortality of living creatures, meaning that the organisms have, for whatever reason, a finite life span. Here, organisms possess a very special mechanism which allows them to reproduce.

If the biological system were aimed exclusively at survival, these two mechanisms would be competing on a par with each other. However, since living creatures also strive to *expand*, the only feasible alternative is the second mode of survival. The regeneration method, while not extinct, is relegated to a subordinate role.

Population growth assumes two different forms. The first one is known as division. This method is associated with, but not limited to, single-cell organisms that split into two parts. However, only the most simple of multicellular organisms, such as weeds, reproduce by division. Evolution has produced an interesting subclass of division-based reproduction where a new organism springs from a small part of its parent. What makes this offshoot technology possible is that each part of an organism contains a program which defines its offspring. This type of reproduction is called *fragmentation*; it occurs in flatworms, sea stars, sea urchins.

The advantage of division as a reproductive method is its economy of means: it utilizes the organism itself as the substrate upon which new organisms are formed. But this method also has its drawbacks. As C. Duddington mentioned,[84] (pp.135-136), in plants, for instance, this mode of reproduction would make it impossible for a new organism to develop at a large distance from its parent, or for the plant to survive under strong fluctuations in temperature (a spore is preserved much more effectively). Generally speaking, division-based methods seem ill-suited for complex organisms when new creatures would have to originate from rather specialized parts of a developed organism. Instead, nature has devised specialized germ cells which develop into a new organism. Over the course of a complex multi-stage process of development which starts out with a single cell, a new multicellular organism is formed.

Because reproduction by division is based entirely on somatic cells let us call this kind of reproduction *somatic*. By analogy, reproduction by means of specialized germ cells will be called *germatic*. Why did I pick this rather unusual term for reproduction based on specialized cells? In the literature on the subject, asexual methods of reproduction are usually opposed to sexual reproduction. A single parent is deemed the distinguishing feature of asexual reproduction which includes division as well as development from a single specialized cell, such as a spore. For my purposes, the crucial criteria in classifying reproductive methods is whether or not reproduction takes place directly using the substance of the parent (somatic cells) or by means of specialized cells, be they germ cells (i.e., cells belonging to two or more sexes) or asexual cells, such as a spore. The term germatic is used to emphasize reproductive modes based on specialized germ cells.

In nature we also observe combinations of two germ cell reproductive methods - sexual and asexual. Ferns are very interesting in this respect.

"The sporophyte, which is the fern plant, reproduces asexually by means of spores, which germinate to produce the prothallus, a tiny plant no more than a quarter of an inch across. The gametophyte bears the sex organs, and the product of its sexual reproduction is the fern plant."[85] (p. 114)

The table below presents different combinations of organism types and methods of reproduction.

**TABLE 4.1.** Organisms having/lacking specialized germ cells.

| Type of organism | Sex | Presence/lack of specialized germ cells | |
|---|---|---|---|
| | | Present | Lacking |
| Single-cell | Asexual | - | Amoebae |
| Multicellular | Asexual | Spores | Many types of weeds |
| | Two sexes | Animals | |

If the sole objective of the living creatures were quantitative growth, reproduction by division would be given top priority, at least for simple multicellular organisms, because it seems to be a more economical and expedient way to procreate - it is simpler and works with the already existing structures.

Once nature had introduced such phenomenon as development and its steadfast companion change, the hierarchy of the various mechanisms of biological dynamics had shifted dramatically. Living creatures must not only survive and grow, but also change in order to become more effective at conquering the environment, with adaptation being one particular case.

An in-depth probe into the somatic and germatic mechanisms of change is discussed in the next chapter. At this point, I would like to note that the somatic mechanism seems more cumbersome, because changes in one set of cells must be transmitted and coordinated with other cells. On the other hand, a germ-cell based mechanism of change allows all changes to take place at one location, using a compact set of elements comprising the specialized cell. Although some features are mechanism-specific, the somatic and germatic mechanisms are not separated by a Chinese Wall. On the contrary, they interact and complement one another.

## 2. EVOLUTION OF THE MECHANISMS OF EVOLUTIONARY CHANGE

Division-based reproduction makes somatic cells the only source of change. Under germ-cell reproduction, two mechanisms of change are possible: a joint germ/somatic cell mechanism and a germ-cell mechanisms only. I shall focus on the first case of joint operation.

### 2.1. Combinations of Germatic and Somatic Mechanisms of Change

With two types of cells, we have the following variables: the vehicle of change can be somatic or germ cells and the source of change can be internal or external. The above parameters can be combined in logical modes either/or as well as and/and. The nine alternatives linking the various mechanisms of change are shown in the following table.

TABLE 4.2. Possible mechanisms of change.

| Cells in which change takes place | Sources of change | | |
|---|---|---|---|
| | External | Internal | Internal and external |
| Somatic | 11 | 12 | 13 |
| Germatic | 21 | 22 | 23 |
| Somatic & Germatic | 31 | 32 | 33 |

I shall list three examples to illustrate the above table. The first reflects combination 11, i.e., change takes place through somatic cells only and results from external factors only. In keeping with the "11" framework , germ cells that partake in reproduction may form as a result of somatic cells producing certain ingredients. This is precisely the view espoused by Charles Darwin in his theory of pangenesis.[86]

The second example reflects the currently prevailing theory of evolution -combination 21 assumes that reproduction is carried out only through the germ cells, and the source of change is primarily the environment.

Third, one can assume that change, which includes both somatic and germ cells, results from a combination of internal and external sources – this is combination 33. I am proponent of this concept, and I shall elaborate it below.

The ruling theory of biological change prior to the 20th century is encapsulated in the first example (combination 11). The 11 camp was not without internal conflicts. For instance, Lamarck and Darwin disagreed as far as the mechanism of change: Lamarck deemed goal-oriented active change possible (the famous example of giraffe's neck), while Darwin emphasized change as adaptation to the environment.

In the 20th century the theory of change exemplified by the second case (combination 21) has become dominant.

## 2.2. Some Hypothetical Examples of the Mechanisms of Change

Let us begin by looking at the general characteristics of the somatic and the germatic mechanisms of change. Some hypothetical examples of the growing developmental complexity of living creatures should clarify the link between these two mechanisms. As far as the sources of change are concerned, in single-cell organisms the somatic and the germ mechanisms of change are one.

Imagine a *two-cell* organism. Each cell is a microcosm, in the sense of being able to divide and produce a new organism. Specialization and exchange of resources improve the performance of both cells. In a two-cell organism, a change in one cell would probably call for appropriate changes in its partner cell. One way for all the needed changes to be implemented is through the transfer of ingredients that would cause change in the partner cell: this is the *aftereffect* method. The process of change can also be accommodated by the mechanism of *prevention*. It entails an iterative process with feedback: the second cell that has undergone change induced by the first cell sends certain cues to the genome of the first cell in order to qualify changes that are taking place in the first cell. In principle, this process might converge to a new well-defined organism.

When change occurs as an aftereffect, there is another way a new two-cell organism might be formed, provided the population as a whole exhibits sufficient diversity. Two-cell organisms, or some subset of organisms, that are part of the population split up into single-cell organisms which then become autonomous (less specialized), with some of them adjusting to some changed cell when they merge to form a new organism.

In the case of a preventive mechanism, the function of change can be implemented via the genome of each pair of cells, meaning that change takes place in the specialized sector of the genome without affecting the more stable sectors of the cell responsible for its daily operations. Following all the mutual adjustments between the changing genomes of the two cells, an organism might actually switch to some new mode of operation.

A transition from a two to a three-cell organism creates new opportunities for change without sacrificing the advantages of specialization. One cell could assume the reproductive functions, i.e., become the germ cell. The problem of origins of the germ cells has a long history. As Claude Villee mentioned[87] (p.12), the current view holds that reproductive cells are formed from non-specialized somatic cells. While one cell could become a germ cell, the other two cells that remain somatic support the germ cell. This scheme opens up new options. For instance, one alternative is for the two somatic cells to reproduce only, thus ensuring that the dead cells are replaced with new ones. Change would be limited to the germ cell. Another alternative is for changes to occur first in the somatic cells and eventually migrate to the germ cell, which will reproduce a changed organism. Another scenario represents a combination of the first and second alternatives, when change takes place in both the somatic and the germ cells.

The above considerations remain valid for multicellular organisms, except that the process of the creation of new organisms involves complex hierarchical structures.

One can also imagine a complex organism in which change is limited only to the somatic cells: the majority of somatic cells support the already existing structures, while some designated subset is involved in innovation. The changing somatic cells coordinate the changes and then gather in one place specially designated for the purpose of producing a new organism.

This kind of somatic mechanism of change might be cumbersome and time-consuming because of the many steps required to harmonize all the changes. The duration of this process hinges on the organism's life-span. The longer the process of innovation transfer among the somatic cells and the shorter the organism's life span, the more vulnerable is the mechanism of somatic change. Its drawbacks would be reduced if we were to relax the assumption that all changes must be completely harmonized with each other or if *partial* changes in the somatic cell were allowed to be transferred to the germ cells. The germ cells, having some special free space in their genome, would slowly accumulate information regarding somatic changes, and this information would somehow be fused with the programs already in the germ cell.

## 2.3. Analogy Between Machine and Organism Design

Before we proceed with our analysis of the advantages of the germ-cell based mechanism of change, especially for complex organisms, I would like to propose an analogy between methods of creation of new organisms and complex machines. The analogy follows the outlined framework regarding the methodology of design of new organisms.

Consider, for example, an engineer who wants to modernize an existing piece of machinery based on the "substrate" of the machine itself. Modernization would be limited to a particular aggregate, with some parts being replaced by new ones. If the to-be-modified aggregate requires no changes in the adjoining aggregate, the problem of modernization is solved. However, in the general case, changes in other blocks are required. The secondary changes to be introduced might not agree with the initially modified aggregate. An engineer will pursue this iterative process in order to attain maximum compatibility between the original aggregate and its co-workers. Eventually this process of integration of all the changes in the various parts of the machine might well improve its overall performance.

The example reveals some problems associated with major modernization based on keeping the fundamental blocks intact. The described method becomes unsuitable altogether in the construction of machines based upon new principles of design and operation; new machines require many new types of aggregate that must be integrated. Of course, new aggregates could be built, regarding this as a sequential multi-stage process, upon relatively universal principles of design modifying them as we go from one stage to another.

As engineering advanced, it became clear that the creation of new machines, as well as major modernization of the old ones, called for a hierarchical process of design based on one informational field, namely the blueprints. Blueprints are nothing but a substrate used to design new models fast and efficiently or to modify and then integrate the units with relative ease.

The key task of such a hierarchical process might be the design of a new aggregate - the heart of the new principle of operation. The general architecture of the machine could then be adjusted to the key unit. Under this approach, it is possible to use standard versatile aggregates that are amenable to modernization. One way to modernize a machine is to use/idle, as the circumstances dictate, certain capabilities of the versatile aggregate.

The hierarchical multi-stage process of design takes advantage of sequential as well as parallel techniques. As new aggregates are designed and various parts modified, inconsistencies might surface. This would call for changes in the blueprints of individual aggregates as well as the overall machine architecture. It would not make sense to have the complete technology of manufacturing the machine from the very beginning. Rather, one ought to focus on the design of individual aggregates and their basic units as well as the technology to implement the next immediate stage. This information would not only provide clues about the ways to proceed at the given stage, but also design and technological guidance for subsequent stages, including ways to alter the information as the process unfolds and the actual results become available.

## 2.4. The Relationship Between the Somatic and the Germatic Mechanisms of Change

The above analogy reveals the advantages of the germ-cell based mechanism of change as compared with the somatic one. In the germ-based mechanism, change takes place at a *holistic* informational level where it is easier to carry out a hierarchical integration of programs governing the development of individual parts of an organism. Under the somatic procedure, the informational profile would change within the individual organs and information would then migrate and be coordinated with the associate organs, finally passing unto the germ cells.

Whatever the structure of the somatic mechanism of change, it is too cumbersome when it comes to restructuring the entire complex organism, and inadequate altogether for organisms based upon new principles of design. On the other hand, a mechanism of change implemented via the germ cells is concentrated in *one place*, so integration of these changes is faster and easier. Moreover, if, as in the previous example, the time required to harmonize the changes is longer than the organism's life span, the intermediate changes can be preserved in the genome and passed on. Apart from housing an established program of development, the genetic structure of a changing germ cell might allocate a special place for storing and creating new programs. This factor is important in understanding the role of heredity in passing on genetic changes. Information that is passed on is not limited to characteristics already expressed in the phenotype, but encompasses newly acquired genetic features as well. I would like to mention here that uncovering these new invisible genetic changes may prove beneficial for an early diagnosis of a pathology.

The somatic mechanism of change, however, should not be ignored when it comes to modifying specific parts of an organism rather than undertaking major restructuring. Here, the somatic mechanism of change may well complement the germatic one, with changes taking place in the organism itself without waiting for a new one to inherit and express the changes, as is the case with the germatic mechanism of change. Whether or not changes in the somatic cells pass on to the progeny, the somatic mechanism complements the germatic one.

Our discussion of the somatic and the germatic mechanisms of change suggests a positive correlation between the role of the germ mechanism and organisms' complexity. Perhaps the somatic mechanism is retained, but a) it complements the germatic mechanism when relatively minor and immediate changes are called for, b) it fulfills specific functions not covered by the germatic mechanism of change, c) it supports the germetic one in order to support such vital evolutionary process as change, and d) finally, it may simply be an anachronism.

If the somatic mechanism of change is an anachronism, it would be similar to an appendix. In herbivorous animals the appendix played an important function. In man its functions are unknown. As long as the appendix is calm, it is not harmful. But, once inflamed, it leads to a pathological condition fraught with death. Outside intervention is required and, if administered on time, pathology can be overcome.

# CHAPTER 5

# INTERNAL MECHANISMS OF CHANGE

## 1. BRIEF COMMENTS ON THE RELATIONSHIP BETWEEN INTERNAL AND EXTERNAL MECHANISMS OF CHANGE

### 1.1. Adaptation, Proadaptation, and Preadaptation

The current reigning doctrine in biology purports that all the different factors external to the genome, regardless of whether the affected changes are induced in somatic or germ cells, produce change in the genome *directly*.

Essentially, external mechanisms are thought to behave randomly via chemical substances, radiation, and similar factors, and semi-ordered operation of these mechanisms is manifest in gene recombination resulting from crossing (assuming, of course, there is some kind of order in selecting one's mate).

The importance of various external factors in driving biological change is beyond doubt. Most theories, other than Lamarckism, assume there is no connection between the actual mutations and the demands of the environment. An interesting theory of so called *adaptive mutations* elaborated in the 1980s tried to break away from this dogma. Its basic thesis is that cells possess mechanisms for selecting appropriate mutations, provided certain preconditions for selecting these mutations are met and no cell growth takes place.[88,89] This theory was vehemently opposed by most biologists. However, the proponents of the "adaptive mutations" theory continued to collect supporting evidence and discovered far-reaching disparities between the mechanism of random mutations and that of adaptive mutations.[90-92] The adaptive mutations hypothesis is still inconclusive, but it warrants further investigation.[93-95]

By generalizing the notion of adaptive mutations, we can distinguish among *adaptive*, *proadaptive*, and *preadaptive* processes.

Adaptive is used denote a process by which an organism prepares itself to respond to changes in the environment whatever these changes may be. This kind of adaptation is predicated on the flexibility of the organs, i.e., their ability to shift from one mode of operation to another, diversity of tools at the organism's disposal (such as different antibodies), etc.

Proadaptation represents a change-specific process by which an organism gets ready for changes in the environment. The currently used biological term, "adaptive mutations", corresponds (my terminology) to proadaptive mutations. The following passage was written by James Shapiro:

> "Neo-Darwinists teach that mutations arise independently of biological needs. Physical insults, chemical fluctuations, and replication errors lead to stochastic changes in DNA sequences. Random mutations mean that the evolutionary watchmaker is blind. To bolster their argument, neo-Darwinists cite an experiment by Luria and Delbruck in which bacterial mutations occur prior to selection for the mutant phenotype. However, this neo-Darwinist position has been challenged for over a decade by the discovery that certain mutations occur much more frequently when they are selected, and thus adaptively useful, than they do during normal growth. The principal difference between these so-called "adaptive mutations" and the phage resistance mutations studied by Luria and Delbruck is that selection for adaptive mutations is not lethal. Thus, nonmutant cells can survive to undergo DNA changes under selective conditions. Not surprisingly, the evolutionary significance of adaptive mutation is highly controversial, and there is great curiosity as to its mechanism..."[96]

Preadaptation commonly refers to "the existence of a prospective function prior to its realization."[97] (p. 86). According to this definition, preadaptation encompasses changes in an organism that were not previously utilized, but which can be utilized in a new environment as well as changes not directly associated with any shifts in the environment.[98] The latter class of changes actually represents the initial stage in a multi-stage process which may eventually bridge these initial stages with changes in the environment. Many,[99,100] but not all, biologists recognize the preadaptive process as it is expounded above.[101]

While the proadaptive process is definitely induced by the environment, adaptive and preadaptive processes are more ambiguous. We cannot rule out that these processes are affected by external factors that are somehow correlated with changes in the genetic program. However, it is more likely that these processes are driven by an internal mechanism of change which supports a dynamic, rather than static, conception of the genome (see section 4 in this chapter).

## 1.2. The Tunnel Process and Preadaptation

An event in an organism that starts from the beginning and fails to serve any purpose in terms of the current conditions is deemed either an atavism, a feature important at some time in the past, or a deformity. However, the problem could be reformulated in the following way: structures without any immediate adaptive purpose are beginning to emerge; the need for them is totally obscure and, in fact, the development of some new structures may never lead directly to any one specific characteristic. The development of these structures may undergo numerous stages of transformation prior to the emergence of a structure which might give rise, relatively quickly, to pragmatically useful characteristics, that is, characteristics that reflect development from the end.

Change from the beginning - the creation of potential for development [1] – seems to have arisen with the emergence of (bio)chemical innovations; however, the origins of this process are clouded in mystery [2]. It seems that changes from the beginning derive not from the morphological creation of potential, but from changes in karyotype, i.e., the set of chromosomes typical of that species in terms of size, number, constituent elements, etc. As Ernst Mayr[102] mentioned, "closely related species often differ more conspicuously in their karyotype than in their morphology."(p. 310)

Over the course of subsequent development, changes in karyotype are expressed morphologically as a new characteristic whose initially obscure purpose emerges, ultimately linking up with the end.

One final comment regarding the tunnel process concerns the relationship between the mechanisms of change and heredity (even if the phenomena are limited to the germ cells). It is essential to take into account the amount of time required for genetic changes initiated at the beginning to reach a stage where the structures generated by these changes can be bridged with the end, in effect, the demands imposed by the environment. Assuming that change from the beginning proceeds rapidly, i.e., these changes link up with the process of change from the end over the reproductive life of the organism in question, the problem of the tunnel process and heredity is resolved within the traditional scheme of change and heredity. However, the time required for beginning-induced changes to prove their worth might be so long that no organism can utilize these changes to its competitive advantage over its reproductive life. Therefore, changes from the beginning will also be passed on genetically, but their purpose in terms of adaptation might come to light only after many generations. This time lag explains the appearance of characteristics that do not seem to conform to the demands imposed by the environment. It is important to keep track of the changes in the genetic structure that manifest themselves only after several generations. Diagnostics based on this kind of hidden long-term changes could be called *very early* diagnostics.

Our discussion of the tunnel process and heredity raises one of the most formidable problems in biology, namely, the emergence of new species via a multi-stage process of change. The most perplexing issue in this field is the possibility that at some intermediate stage of this process a genetic state(s) may emerge that is actually useless or even detrimental to the organism in the context of a given environment. Of course, in the ideal case, any change within an organism aimed at (eventually) creating a new species is also conducive to its adaptation in a given environment. This mode of development, or the process of change, unfolds slowly, and paleontological evidence is expected to corroborate it.

The emergence of a new species, however, does not always follow this mode of evolution. An organism's intermediate states may divert considerable resources for its development to the detriment of the already tested structures, and may actually hinder the organism's hard-earned capacity to adapt to a given environment. This paradigm of development of new species, which parallels the creation of macroevolutionary diversity, raises major but rather poorly explored problems.[103] The most notable exceptions to the rule are the work of Ernst Mayr and the theory of *punctuated equilibria* elaborated by N. Elridge and Steven Gould.[104]

It is no accident that the problem of the emergence of a new species under a multi-stage process of development has been neglected by biologists. Indeed, if one rejects the notion of an internal mechanism of change and the possibility of a tunnel process, it becomes rather agonizing to explain the emergence of unnecessary or even harmful (in terms of a given environment) changes in the genome that are expressed in the phenotype, with all the ensuing (possibly negative) consequences.

Acknowledging the possibility of an internal mechanism of change and the tunnel process opens up new approaches. Changes in the organism aimed at creating a new species may accumulate in some internal structures of the genome and materialize in the phenotype only when these changes are reconciled with the environment. Consequently, the lack of paleontological evidence for intermediate forms heralding the emergence of a new species may be due to the fact that the phenotypes expressing this kind of intermediate change never actually existed. These intermediate changes are also not found in the genes governing the development of a given organism.

Thus, the emergence of a new species via a multi-stage process could also be analyzed in terms of the tunnel process. The implied distinction between changes in the genome and changes in the phenotype extends to changes in the part of the genome connected to the internal mechanism of changes, i.e., the selfish genes.[3]

More detailed examples of the tunnel process may be found in the literature, with many examples that can be analyzed from the standpoint of tunnel process. An interested reader may benefit from learning about one fascinating methodological cue that is presented below.

### 1.3. Origami

Origami is defined in Webster's New World Dictionary as "a traditional Japanese art of folding paper to form flowers, animal figures, etc." The general discussion and the quotes concerning origami are based on a book by Isao Honda[105]. A theoretical analysis of certain aspects of origami is referred to my paper.[106]

I remember my mother's teaching me to construct origami, but I was never taught to invent new origami. The Japanese approach to origami is rooted in an entirely different vision. The child is taught certain basic operations for transforming the initial piece of paper. For traditional origami, these operations are fixed.

"Almost all traditional origami constructions rely on the manipulation of certain regulated and fixed folding methods.

> Creative unregulated folding lines play a part only when the origami process is near completion."(p.27)

> "There are certain origami works that are the accidental result of the addition of a few auxiliary folds to a basic fold during an attempt to develop something else."(p.29)

These random events are still limited, since they are based on the geometric folds.

> "Though there are many other works which develop accidentally when the man folding was actually groping around for some other form, because all of these accidental creations rely on established origami folding lines, their construction is always geometrical." (p.29)

A student of origami first learns how to make certain objects from beginning to end. The next stage is to learn intermediate forms, which in and of themselves do not suggest any final object, but which possess vast potential for subsequent development. Having mastered these lessons, the pupil must then continue on his own. However, the choice of intermediate forms does depend on the final image:

> "...we must first decide what shape we want to make. After various folding attempts we arrive at a number of basic fundamental forms." (p.29)

This approach to constructing origami is not a carbon copy of the biological process of mutation, but perhaps there are some useful parallels: the established folding lines correspond to a fixed number of types of biochemical reactions. The construction of a geometric network of folding lines and intermediate forms representing the potentials for development which give rise to new origami is akin to the development in living creatures of a somewhat ordered network of cells which serves as a foundation for subsequent processes of cell specialization, choice of cell location in the organism, cell interaction, and the creation of intermediate forms which are rather remote from the final organism. One major distinction between origami and biological development is the lack of a preconceived image on the part of the latter, an image that is correlated with the choice of intermediate forms. Perhaps, when it comes to the creation of new living creatures on the basis of an ordered network of cells and intermediate forms, random processes (to some degree) in the internal hierarchy play a more significant role.

In summing up our discussion of external and internal mechanisms of change, we may safely assume that internal and external sources of change interact with each other; it is only natural that there would be some kind of a feedback among programs at different levels. For instance, assuming that the second-level program is linked with the first level one via a feedback mechanism, then an internal mechanism of change would be affected by environmental factors that, even if they have no direct bearing on the second level, do affect the first level program.

Another feature of the interplay between external and internal mechanisms of change is that it allows the process to be not completely ordered. In fact, random perturbations may be conducive to its overall progress. This is not unlike a combination of a completely ordered simplex method with the random Monte Carlo method used in linear programming to solve optimization problems.

In effect, the two sources of change - from the beginning and from the end - should not be opposed to each rather. The only case where this polarization is justified is when all change is end-induced, i.e., when new structures arise to serve some immediate purpose of growth or survival. Assuming that the mechanism of change incorporates different phases/structures of varying proximity to the immediate demands imposed by the environment, the two types of change would seem to compliment each other. The first phase aims at the creation of *predispositions* toward development. In other words, the intermediate states that are selected ought to have the potential to transform into beneficial structures, although there may be no immediate need for them in terms of survival in a given environment. Characteristics that are directly conducive to the survival of the organism/species in a given environment are formed at the second stage.[4] This approach to the relationship between external and internal mechanisms of change suggests an answer to the question posed in the previous chapter in connection with the non-random formation of intermediate genetic combinations over the course of development of an organism having a large number of constituent units (which makes an exhaustive search of all possible variations cumbersome and unmanageable).

I would like to quote an example given by Sydney Brenner regarding the development of worms. Brenner surmises two levels in the genome which control the growth of an organism.

> "The first layer is a noisy, inaccurate set of processes that generates a "sort-of-worm." And the second is a set of refinement processes that tames the unruliness of the first and yields a real and recognizable worm. There is in fact the potential for many types of worm locked up in the genome, but the one that comes out is determined by the refinement genes. Of course, in this process many changes will have 'unpredictable' consequences..."[107]

## 2. PHENOMENOLOGICAL REPRESENTATION OF THE INTERNAL MECHANISM OF CHANGE

In chapter 3, I pointed out two ways to represent a system via a law (analytically) or via a mechanism that coordinates the system's constituent elements, subject to certain constraints on the principles of their interaction.

Biologists conducting phenomenological research into organisms' internal mechanisms of change have been guided by the traditional search of natural scientists for *laws* with the organism itself, regarded as a black box. These laws pertain to such variables as the shifts in the ratio between different mutations, as well as the development of specific mutations or even resulting individual organs.

It has been verified at the phenomenological level that there is a pattern to the distribution of random mutations[108].

The laws allegedly governing the evolution of individual organs will be discussed below.

The reader should now to refer back to Table 3.2 (Chapter 3). *Law-based* theories of biological change, which Ernst Mayr would call *telenomic*, are not homogeneous. Some of them are intertwined in a strange manner with classical Darwinism, in the sense that they presume change from the end, using the tunnel process framework, meaning the demands imposed by the environment.

According to Ludwig Doderlein,[109] the initial development of a species follows the classical struggle for survival: a species asserts itself in the world by accumulating changes in the various organs; subsequently, these changes are passed on and reinforced/preserved through the process of selection. However, this mode of development changes once the species has displaced its competition. The changed organs which have ensured survival by utilizing the experience of previous generations continue to change by themselves as a result of inertia.[5] At first, such changes might be merely useless, but eventually they become detrimental and lead to the demise of the species. Doderleine illustrates his theory with such examples as the saber toothed tigers, mammoths, giant deer, and *Babirussa* boars living on the island of Celebes. Once this species of boars was free of competition, their fangs continued to grow, in some cases penetrating the scull. Doderleine's theory can be summarized as follows: the evolution of a species follows a bell-shaped curve.

Leo Berg produced a variation on Doderleine's evolutionary theme.[110] Berg's concept introduces an explicit assumption that changes evolve from the beginning, meaning that changes unfold according to some internal laws that are independent of the environment. Moreover, these laws are not necessarily driven by inertia. Berg illustrated his concept with numerous examples from the plant and animal kingdoms.

One of the most interesting schemes belonging to this class of "law-based" theories of evolution is the concept of aristogenesis advanced by Henry Osborn.[111] It claims that certain aristogenes, whose function is rather obscure, appear in the germ cells. As aristogenes evolve, they generate very important organs. Osborn ascribed the emergence of aristogenes to environment-induced factors. Later, he claimed, that aristogenes develop on their own accord.

What fascinated me about Osborn's concept was the idea that, not only is change initiated at the beginning, but the purpose of the various structures formed at these early stages is not clear. Therefore, while the environment does exert some obscure pressure upon the formation of aristogenes, there is no direct program-like connection between the final demands of the environment and the initial changes. Also, once aristogenes are formed, they evolve on their own.

Using my terminology, aristogenes could be compared to potentials, which are capable of developing in different directions within some feasible space. In fact, as potentials evolve based on some internal principles of development, they reach a transition phase, at which point they might actually begin to assimilate signals from the environment.

## 3. STRUCTURAL REPRESENTATION OF THE INTERNAL MECHANISM OF CHANGE

### 3.1. Preliminary remarks

Most biologists reject the notion of an internal genetic mechanism of change. This controversy has a long history. It seems to me that the aversion on the part of the biological community toward research into the internal mechanisms of change is due to the fact that their own investigations, fueled by the rapid technological advances in molecular biology, are focused on the study and search for genes responsible for the development, operation, and interaction of specific organs. In addition, they seek to uncover flaws in the mechanisms regulating these genes - flaws that give rise to certain pathologies such as cancer.

A number of biologists, however, have advocated a different approach to the genome as a dynamic system - a thesis which runs contrary to the prevailing view of a genome as a static system subject to random external fluctuations.

One of the leading scholars in this group is James Shapiro. He very early was aware of the importance of McClintock's work.[112] Shapiro constructed a theoretical as well as an experimental framework for the concept of the genetic system as an information-processing dynamic system rather than a mechano-chemical one. In his article[113], he even called genomes "smart systems."

> "Our thinking about genetic systems needs to become much more sophisticated before we will be in a position to comprehend the contents of massive DNA sequence data bases and use the information effectively. Instead of the basically mechano-chemical view of the genome inherited from neo-Darwinism and the early years of molecular biology, we need to think of genomes as information-processing systems."

In the same article, Shapiro also underscored the existence of non-traditional mechanisms responsible for restructuring the genome (beside the well-known ones such as damage repair).

> "Complementary studies have examined how genetic change comes about, either in germ line, through the action of

transposable elements or recombination and mutator functions, or in somatic tissues during developmental DNA rearrangements. The results show quite clearly that many (perhaps the vast majority) of DNA alterations are not due to chance chemical events or replication errors. Rather, they result from the action of highly sophisticated biochemical systems which may be thought of as genome reprogramming functions."[6]

My working assumption also recognizes the existence of an internal mechanism of change.

The key to my approach to biological change is recognizing the importance of the internal mechanisms of change which I assume are based on the existence of a program within the genetic structure which affects programs of development regardless of whether the affected changes are occuring, in somatic or germ cells, and whether their are induced endogenously or exogeneously.

Current research in developmental biology is slowly becoming receptive to the idea of genetic programs that alter lower level genetic program.[114] This line of research is also being pursued by biologists at the molecular level. For instance, whereas previously research was focused on the genes responsible for the formation and interaction of the various organs - first-level program - in more recent years, the emphasis has shifted toward the second-level program which involves transposons, etc. All this basically conforms to the concept of a dynamic genome introduced by Barbara McClintock[115]

Molecular research into the higher order genetic programs that affect the code does not repudiate classical biology, with its emphasis upon the phenomenological course of development of different species. The two branches complement each other, but fusing them into a unified vision still must be accomplished.[116]

According to the tunnel process, internal mechanisms of change can proceed from the end as well as from the beginning. In the first case, the program-changing program in the genome is triggered by external factors; and, in the latter case, the program-changing program is governed by its own internal principles.

Having programs at just two levels, i.e., organism-structuring program and program-structuring program, suggests the autonomous workings of the internal mechanism of change. The second-level program can modify the first level program and if the genome possesses a third level program. There is even more room for change due to internal causes because the first level program is potentially affected by the third and the second level programs.

We can safely assume that internal and external sources of change interact with each other: it is only natural that there should be some kind of a *feedback* among programs at different levels. For instance, assuming that the second-level program is linked with the first level one via a feedback mechanism, an internal mechanism of change would be affected by environmental factors that, even if they have no direct bearing on the second level, do affect the first level program.

Although these assertions regarding an internal mechanism of change are hypothetical and unconfirmed, there is evidence that attests to the existence of the necessary preconditions, including:

1) pronounced predominance in the genome of selfish genes (also called *junk* or *trash* DNA) whose function is unknown (so called C-value paradox) and which probably partake of the internal mechanism of change;

2) certain regularity exhibited by the selfish genes;

3) the presence of transposons that partake directly in the process of change;

4) the presence in the genome of a powerful mechanism of ordered search through numerous possible combinations.

## 3.2. The C Value Paradox

What is the so called C-value paradox? Essentially, it states that the amount of DNA in the genome is much greater than the amount needed to code for the various proteins which form an organism.

> "The **C value paradox** takes it name from our inability to account for the content of the genome in terms of known functions. It expresses the existence of two puzzling features:
> There may be large variations in C values [the total amount of DNA in the (haploid) genom of each living species] between certain species whose apparent complexity does not vary much. In amphibians, the smallest genomes are just below $10^9$ bp, while the largest are almost $10^{11}$ bp. It is hard to believe that this could reflect a 100-fold variation in the number of genes needed to specify different amphibians.

> To reinforce this skepticism, some closely related species show surprising variations in total genome size. For example, two amphibian species whose overall morphologies are very similar may have a difference of, say, 10 in their relative amounts of DNA. It seems unlikely that there could be a tenfold difference in their gene number. Yet if the gene number is roughly similar, most of the DNA in the species with the larger genome cannot be concerned with coding for protein: what can be its function?
>
> There is an apparent excess of DNA compared with the amount that could be expected to code for proteins. Indeed, this is often referred to as the problem of excess eukaryotic DNA. We now know that some of the excess is accounted for because genes are much larger than the sequences needed to code for proteins (principally because of the intervening sequences that may break up a coding region into different segments). We do not know yet whether this form of organization is sufficient to resolve the problem."[117] (p. 67)

Research into selfish genes seems to confirm that some of these genes are involved in genome repair or regulation,[118] but the nature of this class of genes remains a mystery.

One possible scenario underlying the C value paradox is that selfish genes are employed in the internal transformation process that generate new organisms.[7] One indirect confirmation of this hypothesis is the varying size of the genome of different organisms. It is interesting to note that the largest genome belongs to plants and amphibians (up to $10^{11}$), rather than complex mammals whose genome is of the order of $10^9$. In fact,

> "Mammals have genomes that fall into a particularly small range of DNA contents, with a C value usually of 2-3 picograms ($2-3 \times 10^9$ base pairs). Amphibia, by contrast, vary very much more widely, from less than 1 picogram to almost 100 picograms. Even closely related amphibia may have greatly different content of DNA in the haploid genome."[119] (p. 88)

This gap in the C value bears an intriguing correlation with completely independent statements by biologists regarding the origins of the species.

> "New species do not evolve from the most advanced and specialized forms already living, but from relatively simple, unspecialized forms. Thus mammals did not evolve from the large, specialized dinosaurus but from a group of rather small and unspecialized reptiles."[120] (p. 751)

It is quite plausible that this correlation reflects some kind of a causal connection. It would seem that the greater the C value (other conditions being equal), the more room there is for radical changes in an organism, including changes that lead to the birth of new species. However, "room for change" is not a sufficient condition. It is necessary to have sufficient variety of building blocks and biochemical processes to ensure subsequent dynamics within this space. A very simple organism, even one whose C value is high, has relatively little opportunity to develop. It is deficient in the number of ready blocks-organs and lacks sufficient variety of biochemical processes, particularly their combinations, that all contribute to the process of creation. On the other hand, an organism that is overly complex is not too amenable to change, because change is limited by all the different combinations of the already existing block-organs. It follows that these intermediate (complexity wise) organisms, such as amphibians, have the greatest opportunity to change.[8]

The C value paradox is unlikely to be resolved within biology's traditional frame of reference. Most mainstream biologists refer to excess genes as junk, trash, etc. - terms that reflect the negative attitude of biologists toward the role of these components in the process of evolution.

Richard Dawkins was one of the first scientists to attempt to link the function of these excessive genes with their evolutionary significance.[121]

He put forth a definite concept of selfish genes in which the unit of evolutionary selection was not an organism but a gene, i.e., selection bears directly upon the genes, and their struggle for survival is essential in and of itself. The question then arises regarding the link between the autonomous behavior of the genes and the development of the organism's phenotype, or at least the host cell in which these genes reside.

According to one rather extreme point of view

> "When a given DNA, or class of DNAs, of unproved phenotypic function can be shown to have evolved a strategy (such as transposition) which ensures its genomic survival,

then no other explanation for its existence is necessary. The search for other explanations may prove, if not intellectually sterile, ultimately futile."[122] (pp. 601-603).

Less radical opinion purports to reconcile the autonomous evolutionary development of the genome with some beneficial function fulfilled by these excessive genes (by analogy, even moderately harmful parasites can form a symbiosis with the host organism). This autonomous development of the genome might become detrimental if it exceeds a certain threshold by diverting too many nutrients and energies into sustaining the genome itself. Excessive uncontrolled expansion of the genome is sometimes compared with *genome cancer*.[123]

Even biologists who endeavored to look for the functions of these unknown genes have managed to come up with what are, at best, only partial hints inspired by the prevailing dogma.

In 1992, for example, a group of European scientists made outstanding discoveries in decoding the structure of the DNA. They were able to complete a project aimed at uncovering the sequence of a complete eukaryote chromosome: the 315,357 base pairs of chromosome III of the yeast Saccharomyces cervisiae. Along the way, they uncovered an entire class of genes which comprised more than half the chromosom and whose function is completely obscure. In generalizing the results, the author of the project report wrote: "Searching for the function of the new genes is going to be a time consuming business-far tougher than the original sequencing."[124]

Project coordinator, Piotr Slonimsky, who was responsible for the functional analysis of the data, lists a number of specific factors which might trigger the functions of these functionally mysterious genes – for instance, morphology, temperature sensitivity, etc. However, there is no allusion to the idea that these genes might partake of the internal mechanism of evolution.

## 4. STRUCTURAL REPRESENTATION OF THE INTERNAL MECHANISM OF CHANGE (CONTINUATION)

### 4.1. Patterns exhibited by selfish DNA

It has been established that the genome is comprised of predominantly selfish DNA whose role is unclear. Assuming that these sequences embody the internal mechanism of change, they must be organized.

The last few years have witnessed the emergence of strong evidence attesting to the orderly nature of this class of genes. A group of Boston scientists have uncovered a linguistic pattern in the selfish genes.[125] Using sophisticated techniques borrowed from linguistics, ranking the words in order of frequency, text redundancy, these scholars "have shown fairly clearly that the 'junk' has all the features of a language". Interestingly, the same tests applied to familiar coding segments with known functions determined that these parts lacked linguistic features. My interpretation of this discovery is the following: since coding segments represent a set of instructions, like all instructions, they make insipid the creative features (as outlined above) of a living language that are generally directed at serving creative purposes.

The Boston group in its article concludes that whatever the function of these selfish genes is, the discovered regularity signifies that these genes do send meaningful messages:

> "We extend the Zipf approach to analyzing linguistic texts to the statistical study of DNA base pair sequences and find that the noncoding regions are more similar to natural languages than the coding regions. We also adapt the Shannon approach to quantifying the "redundancy" of a linguistic text in terms of a measurable entropy function, and demonstrate that noncoding regions in eukaryotes display a smaller entropy and larger redundancy than coding regions, supporting the possibility that noncoding regions of DNA may carry biological information."[126]

In fact, new data suggests that the messages produced by selfish genes pertain to the process of change which does not exclude other regulatory or repair functions of these genes.

## 4.2. Transposons

I believe that the starting point for any discussion concerning the role of the selfish genes should be Barbara McClintock's research into transposons.[127] McClintock's research in corn mutations demonstrated the presence of an internal mechanism of change which allows transposable genes - transposons (first called *jumping genes*, and then *mobile genes*) and multi-factor biochemical systems to reconstruct DNA molecules.[128] Reconstruction takes place at the chromosome level, meaning that the jumping genes had transposed from one chromosome to another. This transposition was "fixed" along the chromosome, thus passing on to the next generation. McClintock began to explore the jumping genes phenomenon as early as 1944, and that time she was the sole advocate of this idea.[129] Public accounts of her jumping-genes concept date to the early 1950s and in the early 1960s the existence of these genetic components was rigorously documented. More advanced work focusing on the molecular structure of these elements began in the mid-1970s.[130]

From its very inception, McClintock's concept of jumping genes was rejected by most biologists who opposed the very idea of an internal mechanism of change. In spite of the fact that by the time her ideas became public McClintock was a well-known and respected geneticist (in 1939 she was elected Vice President of the Genetics Society of America, in 1944 she became a member of the National Academy of Sciences, and in 1945 the President of the Genetics Society), her reputation among her colleagues was that of a "crazy old woman." There were many reasons why McClintock's revolutionary ideas ran against the grain of conventional biological theory: First and foremost was the discovery of DNA in the 1940-50s as the vehicle of genetic information, the double helix structure of the DNA, the launching of the central dogma of molecular biology (purporting a rigid sequence: DNA-RNA-proteins, precluding any feedback), shift in the target of experimental molecular biology from multicellular organisms (McClintock's corn experiments) to a single cell, as well as many other reasons. However, following the initial excitement over all these great discoveries, there was a renewed interest in exploring McClintock's ideas, now at the molecular level. This renewed fascination was sufficient for McClintock to be awarded the Nobel prize in 1983 for "Discovery of mobile genes in the chromosomes of a plant that change the future generations of plants they produce."

However, this was not a happy end to an otherwise rocky road, but only an *intermediate event*. While recognized by some biologists, including James Shapiro, McClintock's ideas did not receive universal acceptance in the biological community. Even today, many biologists have a poor understanding of her achievements. In 1983 Evelyn Fox Keller wrote in her book devoted to the life and work of Barbara McClintock: "Barbara McClintock remains in crucial respect an outsider."[131]( p.xii). Little has changed since then as far as the attitude espoused by the leading biologists toward McClintock. For instance, in 1990 a conference was held devoted to new trends in evolutionary theory. It brought together many leading American evolutionists. The name of Barbara McClintock was missing. It was not to be found in the Index of names listed in the conference proceedings.[132] The sole exception was a paper devoted to the behavior of jumping genes in *Drosophila*.[133] The paper mentions in the introduction that the development of the concept of the dynamic genome based on mobile elements (as opposed to the prevailing static concept) ranks among the most important recent discoveries in genetics. The authors pay tribute to the work <u>Mobile Genetic Elements</u>, edited by J. Shapiro. There is a single sentence to the effect that about 40 years earlier Barbara McClintock discovered transposable elements and that this discovery was met with great skepticism, and it was only in the last two decades that these elements were found in abundance in living organisms. No reference was made to any of McClintock's work.

While most evolutionists are somewhat condescending to ward McClintock's ideas, a number of molecular biologists are beginning to recognize her achievements.

In his obituary statement James Shapiro wrote the following words about McClintock: "One day, she may well be seen as the key figure in the 20th century biology."[134]

His praise is based not only on McClintock's discovery of mobil elements, but also on her global vision of the biological mechanisms of change, vision which might prove influential in the 21 century.

> "...standard theories are still framed in terms of independent genetic units, whereas McClintock thought of the genome as a complex unified system exquisitely integrated into the cell and the organism.
> ...There is good reason to believe that McClintock's integral view of the genome will prove to be prophetic."

At the same time, a number of biologists continue the work on transposons conducted by Barbara McClintock, endorsing the idea that the selfish genes are the source for transposons and, in general, a vehicle of an internal mechanism of change.

John McDonald organized a conference on "Transposable Elements and Evolution" held on June 27 and 28, 1992, at the University of Georgia in Athens.

Molecular biologist Rene Herrera of Florida International University in Miami spoke at the conference:

> "The concept of useless DNA is now obsolete", [adding *A.K.*] that the selfish DNA theory, which holds that the only function of the elements is to reproduce themselves and spread throughout the genome, might have been acceptable a few years ago when little was known about transposable elements, but is no longer."[135]

Claims were made at this conference, corroborated by preliminary experimental data, to the effect that transposable genes can explain changes not only within a single organism, but also the origins of new species. Moreover, these genes are able to function outside a given organism, moving from one species to another.

The work of Richard M. von Sternberg *et al.*[136] elaborated upon some very interesting results regarding the inner mechanism of change in connection with transposable genes and their interaction with Interspersed Repetitive Elements.

I am enthusiatic about all of the discoveries linked with internal evolutionary mechanisms of change. At the same time, I feel the greatest limitation of research in this area is the small number of specific types of internal mechanisms which have been shown to cause change. Only one such mechanism, namely mobile genes, has been studied in depth. The champions of this mechanism try to use it to explain as broad a spectrum of changes in the organism as possible (as seemed to be the case at the aforementioned Meeting on "Transposable Elements and Evolution").

## 5. STRUCTURAL REPRESENTATION OF THE INTERNAL MECHANISM OF CHANGE: COMPUTER BASED ON DNA-MOLECULE

Assuming that selfish genes incorporated into DNA and the transposons partake of the internal mechanism of change, then one precondition for this mechanism to perform successfully is that the structure of the DNA molecule must incorporate an elaborate mechanism of searching through and selecting from the myriad of possible combinations of constituent elements. In other words, an internal mechanism of change signifies that the genome is not merely a set of instructions for coding RNA and respective proteins; the mechanism must also perform recombinations on the elements comprising the genome. In fact, it could be stipulated that, in the event the initial structure of the DNA is somehow ordered, the mechanism also generates ordered combinations.

The computer, recently invented by an outstanding American scientist Leonard Adleman, based on the DNA molecule confirms unequivocally the presence of a genetic mechanism of selection[137]. Edelman proved experimentally that the computer he created has the capacity to solve the so called *seven cities* problem. (In effect, there are seven cities interlinked via one-way or two-way routes. Determine whether it is possible to travel from any one city to any other city without traversing the same path.) This important problem, known as the directed Hamilton path problem, embraces many different problems including the search for optimal solutions, some logical problems (Boolean logic satisfaction), and basically all search problems where the number of combinations grows exponentially with the number of variables.[138]

Edelman's computer could be classified as an analog device, since the computations are performed using natural processes, in this case, biochemical ones rather than some artificial informational program. (In principle, the criteria distinguishing between digital and analog devices are whether the operations are discreet or continuous). The data are entered by using techniques of genetic engineering to form the required sequence of basis amino acids. The solution is also represented by the resulting configuration of the DNA molecule.

In comparison with conventional computers, the performance of the DNA computer is astounding; its advantage in speed is abillion-fold, while its energy and space requirements trillions of times less. While each individual operation performed by the DNA machine takes considerably more time than its conventional counterpart, the fantastic speed is achieved via an incredibly high degree of parallelism. Edelman's computer is not universally superior for all applications. It suffers from numerous drawbacks, and only more research will determine its scope of application.[139]

The important point in terms of our discussion is that Adleman's computer is an experimental verification of the genome's incredible propensity for ordered processing of information. Its implications for biology are best summarized by David Gifford:

> "For biologists, the results indicate that simple biological systems have the ability to compute in unexpected ways. Such new computational models for biological systems could have implications for the mechanisms that underlie such important biological systems as evolution and immune system".[140]

A number of biologists choosing not to ignore "Adleman's computer" have essentially acknowledged that the genome does incorporate a performance mechanism, implying that it is more than just a set of DNA coding instructions. Still clinging to the prevailing notion of the random nature of mutations, however they insist that the genome operates in a random mode.

Let me quote Kevin Kelly known for his work on the applications of new biology to machines, social systems, and economics.

> "When we understand that computation is a very broad term that includes a range of calculations as different as the way a test tube of DNA finds a sequence of letters or the way an adding machine tallies a sum, then we can see something else: the DNA in our cells also computes. Over thousands of generations, the DNA in human cells can produce a new body-and-mind form that will be better adapted to our environment. It does this by trying a billion different versions all at once. You are one possible answer. So am I. This may not be the smartest way to modify a design, but then again, we know it works."[141]

In summary, a definitive internal mechanism of change has not been established but numerous discoveries in biology have laid a firm foundation for the possibility of its existence. Positive evidence includes: the predominance in the genome of order-exhibiting surplus genes with unknown functions (so called C value paradox); a linguistic pattern in the selfish genes; the presence of transposons among the selfish genes; and, finally, a powerful mechanism of search incorporated into the genome.

One other benefit of hypothesizing an internal mechanism of change is to become open to *assimilate* new discoveries in molecular biology pertaining to selfish genes. All too often, proponents of the traditional approach, especially evolutionary biologists, tend to deny or simply ignore these discoveries, while the less militant ones mold new data to fit it into the orthodox framework.

## NOTES TO CHAPTER 5

[1]. The term *evolutionary potentials* does appear in the literature. In a sense it is synonymous with the generic term *potential for development*: "one has to admit that the immense comprehensiveness of the hierarchy of life proves that the first organisms that occurred on the globe must have been provided with great developmental possibilities. An adequate term for these possibilities is evidently "evolutionary potentials"."[142]

[2]. "Chemical inventions on the cellular level are the prerequisite of some of the most important adaptive shifts. Alas, our knowledge of comparative biochemistry is still far too rudimentary to tell us whether or not it was a biochemical invention that gave the mollusks, crustaceans, and other now dominant groups of marine invertebrates their ascendancy over eurypterids, trilobites, graptolites, and brachiopods, once the rulers of the seas."[143] p. 62.

[3]. Similar ideas (unrelated to the tunnel process scheme) regarding the emergence of new species were expounded by S. Gould and E. Vrba: "Perhaps repeated copies (of DNA, *A.K.*) can originate for no adaptive reason that concerns the traditional Darwinian level of phenotypic advantage.... Some DNA elements are transposable, if they can duplicate and move, what is to stop their accumulation as long as they remain invisible the phenotype (if they become so numerous that they begin to exert an energetic constraint then natural selection will eliminate them)? Such "selfish DNA" may be playing its own evolutionary game at a genic level, but it represents a true nonadaptation at the level of the phenotype. Thus, repeated DNA may often arise as a nonadaptation. Such a statement in no way argues against its vital importance for evolutionary futures [that is, for subsequent phenotypical adaptation of repeated DNA]. When used to great advantage in that future, these repeated copies are exaptations."[144]

*Note.* "Exaptation is Gould's neologism, one introduced to distinguish structures whose current functional role is not part of the explanation of their origin from those whose roles are."[145] (p. 244)

[4]. By analogy, this pair of biological structures might be compared to the left and the right hemispheres of the brain. The latter forms a artistic image which creates predisposition to a solution; the left side is responsible for rigid logical manipulations.

[5]. A particular case of Doderleine's general theory is the concept elaborated by Othenoi Abel.[146] The only recognized source of changes in living creatures is inertia. This constitutes the so called "Abel's Law", deduced from the assumption that the organic world is based on the laws of physics, which includes mechanics and the law of inertia.

[6]. J. Shapiro in his note reiterated his approach to the internal mechanism of change. "Molecular genetics has revolutionized our understanding of cellular mutational

mechanisms. The new information alters some of our underlying assumptions. On the one hand, we now know that elaborate repair regimes take care of accidental genomic damage (for example, radiation and chemical insults, replication errors). Thus, these random events diminish as potential sources of evolutionary variation. On the other hand, it is now clear that cells contain multiple, sophisticated, natural genetic engineering systems (nucleases, ligases, topoisomerases, recombinases, transposons, retrotransposons, plasmids, viruses), and we increasingly appreciate these cellular biochemical activities as important mutagenic agents. The versatile operations of these systems include insertion, deletion, inversion, fusion, amplification, dispersed and tandem reiteration, and other DNA rearrangements.

... I find it more reasonable to think of mutational events (which may involve many precise biochemical reactions) as resulting from the concerted action of dedicated cellular machines than as accidents or "pathologies." My argument is that these "high-tech" natural genetic engineering systems serve an adaptive function by generating the hereditary variability needed for short and long-term survival. They provide the biochemical activities that account for evolutionary patterns of genome organization unanticipated by conventional theory shuffling of sequences encoding protein domains, assembly of regulatory regions containing multiple transcription factor binding sites, duplication and dispersal of sequences among gene families, and amplification of repetitive DNA elements."[147]

[7]. In a developed society the proportion of people involved in R&D sector increases dramatically. Just compare a primitive society in which all members are involved in routine tasks with a modern society where new ideas and their implementation draw upon people in numerous research organizations.

[8]. This approach to development bears an affinity to origami. When its structure is extremely simple, initially, a flat piece of paper, there is plenty of potential room for development. However, it is impossible to produce complex origami at the initial stages of the process because of a lack of intermediate structures as well as allowable operations upon these structures; for instance, one cannot use twisting out until certain intermediate structures amenable to this operation are created. Once a complex origami is created, change is limited by the large number of rather rigidly interconnected constituent parts. It turns out, that the greatest room for creativity is at the intermediate stages. Here, we have a sufficient variety of units which allows many different transformations to be performed upon them.

CHAPTER 6

# SPECIAL FEATURES OF THE SOMATIC MECHANISM OF CHANGE

This chapter presents a broader picture of the somatic mechanism of change outlined in previous chapters. In going through this chapter, the reader should keep the following things in perspective: 1) The great majority of biologists today reject outright the notion of the somatic mechanism of change. As a result, literature that might examine the subject in any sort of systematic fashion is lacking. Therefore, our subsequent discussion, although corroborated by some empirical evidence, is primarily the fruit of my own deliberations; 2) Almost every stage of the hypothesized process of somatic change is inconclusive, since it is postulated on comparing and contrasting it to the development of cancer; 3) My aim was to outline a general approach to the mechanism of biological change, therefore, the discussion of each stage of somatic change is brief; and 4) I am not claiming that the changes induced by the somatic mechanism of change are transferred to the germ cells; for my purposes, the mere existence of the somatic mechanism of change is sufficient. If it is determined that somatic cells exert influence over germ cells, that would make the pathological expression of somatic change ever more complex, if only because these changes are passed on to future generations. With all these qualifications, my discussion of the somatic mechanism, is rather sketchy.

I shall begin by outlining the stages of the somatic mechanism of change and then proceed to discuss each stage individually.

## 1. THE STAGES OF THE PROCESS OF SOMATIC CHANGE

One general characteristic of the mechanism of somatic change is that it represents a multi-stage process. It incorporates a wide array of tools responsible for changing the genetic program that forms an organism as well as for regulating the pace and depth of these changes. This change is carried out primarily by means of horizontal mechanisms, i.e., via interactive behavior of changing cells.

The process of somatic change may consist of the following stages:

1. When considered from different perspectives the sources of change may be quite diverse:

The temporal aspect, meaning at which stage of the organism's development (starting with the embryonic stage) do changes take place;

The hierarchical aspect - changes that take place in the organism range from changes in individual genes to changes in the organism as a whole;

The genesis of the sources of change which range from external factors such as chemicals, radiation, and viruses, to self-induced internal processes.

2. The more critical the change, i.e., the more sectors of the genetic code that are affected, the less differentiated the cell must be in order to free itself of the forces hindering its diverse genetic capacities. It is quite plausible that a cell undergoing a major change would initially act as if it were degenerating (negative differentiation) or that less mature, and thus less differentiated cells, would be involved in the process of change. These features are characteristic of the *stem cells* that have the capacity to develop into different specialized cells (within certain limits). For instance, in the blood producing system of mammals, stem cells can develop into erythrocytes, trombocytes or leukocytes. The versatility of the stem cells varies from unipotent cells, which can develop into only one type of differentiated cells, to oligopotent cells that can evolve into a few types of differentiated cells, and to pluripotent cells, that are capable of developing into many different types of differentiated cells. Moreover, stem cells can grow unchecked, opting to remain stem cells or become irreversibly differentiated.[148,149]

3. A less differentiated cell behaves according to changes in its genetic structure: within the existing genome by activating the suppressed genes/suppressing active genes, or by incorporating new components in the genome (this takes place when a virus enters the genome), or through cell fusion (mentioned in connection with Paramecium reproduction).

4. Less differentiation is also related to the need on the part of an innovator cell to be independent economically from the governing mechanism. Economic autonomy allows the innovator cells to acquire nutrients by bypassing the controls imposed by the established mechanism

of resource allocation. To achieve this end, the innovator cell must simplify its metabolism. It can be done by, switching from oxygen metabolism, characteristic of highly developed cells, to primarily anaerobic, or by reducing the number of receptors linking it with other cells, etc. Under certain conditions, the reverse tactic is exercised, i.e., when an innovator cell augments its link with the organism (still preserving its high degree of differentiation and a complex system of oxygen metabolism) by increasing the number of receptors, by greater mutability (assuming some mutations are more readily adopted by the organism than the existing ones), etc. Augmented ties may make it easier for the cell (taking into account its population size) to acquire the additional nutrients it needs for rapid development.

5. Changes in the cells vary in scale, i.e., the scope of changes is very wide. A rough classification gives us three arbitrary categories: minor, intermediate, and major changes.

By definition, a cell which has undergone minor change is able to *localize* its development to a given organ. This typically occurs when adaptation to changes in the environment calls for temporary, not inherited, changes in cell operation. (See in Chapter 3 the role of the viruses in an organism's adaptation to changing conditions.)

Intermediate changes mean that the cell must be capable of *interacting* with other cells thus paving the way for the creation of a new organ or changes in the existing organs.

A cell with major innovative tendencies must be capable of coordinating the proposed changes with cells in other organs. Eventually all of the information must find its way into the germ cells thus passing on to the progeny.

6. At the early stages of cell development (M1 and M2), changes in the cell are linked to different mechanisms, particularly to telomeres, since telomeres also act to preserve chromosome structure (more about telomeres in section 2 of this Chapter).

7. Each DNA type possesses its own damage-repair mechanism that solves very elegantly problems which, at first glance, seem insurmountable; in a sense, this mechanism acts to prevent sporadic breakdowns.[1] Presumably, changes of an innovative nature need not trigger the repair mechanism which either fails to react to innovations or possesses genes that block its actions. The existence of such a selective mechanism of repair is indirectly confirmed by different segments of the gene having a different propensity for repair.[150]

One theory claims that the death of a cell is also preprogrammed, meaning that cell interaction activates some internal mechanism that leads to a kind of suicide (*apoptosis*).[151] A recent scientific discovery found two genes in Drosophile (ced-3, ced-4) whose joint action results in the death of a normal cell.

It is well known that apoptosis normally surfaces during the course of embryonic development. Perhaps apoptosis takes place when the incurred damage cannot be repaired. In other words, the process of apoptosis may weed out poorly mutated cells, including those that may pose a threat to the organism.[2]

8. Following its renewed specialization, the cell must resume its reproductive cycle. Telomere recovery is crucial, since a cell that has exhausted its telomeres may not be able to reproduce.

9. The changed cell must reproduce faster in order to have time to influence the development of its host organ or other organs and, in the case of germ cell reproduction, possibly even the germ cells.

10. The mechanism of accelerated cell growth suggests the presence of its counterpart - a mechanism that impedes cell growth. This widespread duality is due to the fact that, to get the process of accelerated growth started, more effort (greater pace) is required than is necessary for the process to continue at its regular pace. Brakes are used to slow the process to its normal pace.[3]

The discovery of gene *p.16* by D.Beach confirms that normal cells possess such a suppressor gene – a gene that halts cell multiplication and causes the cell to revert to a calm state.[152]

11. As changed cells develop, they may form new structures that are precursors of new organs; of course, the degree of novelty will vary. These new formations may require an infrastructure, including the blood supply system, to ensure the necessary supply of nutrients to its cells. Organisms possess an intricate mechanism to ensure the creation of such an infrastructure in the tissue. This mechanism incorporates growth factors (fibroblast growth factor and VEGF) as well as inhibitors (trombospondin and platelets). Heparin, in turn, controls both the growth factors of the new blood supply system and the inhibitors.[153]

12. One way for a cell to convey information to other cells is to excrete respective genetic information; the other mode is for the entire cell to penetrate other organs.

In order for a cell to travel from one organ to another, there must be a mechanism of cessation from the living tissue of the host organ. Such a mechanism is part of the genetic code, i.e., the code contains genes governing cell separation from the tissue.[154]

13. One might assume a certain regularity or pattern in cell migration. Perhaps the changed cells spread throughout the organism, but they take root primarily in those organs that are somehow related. Perhaps they consume/supply nutrients to the target organs or are morphologically related to them. Finally, they may be linked to organs that are formed sequentially as the genetic program unfolds.

Another plausible scenario is that the process of coordination in the hereditary somatic mechanism is of iterative nature, meaning that the changed cells from a given organ that have traveled to other organs come back to the original organ more informed.

14. Political independence of an innovator cell means that the immune system should not put excessive pressure on it. One can also assume that an immune system operating under the conditions of equilibrium would be too rigid, suppressing or preventing dissident cells from taking root. It is quite plausible that the immune system itself possesses a mechanism that weakens it somewhat (but only somewhat) in order to allow new entities to develop. This mechanism is probably turned on as the process of change unfolds.

## 2. SOME COMMENTS ON CELL DIVISION AND THE PROCESS OF CHANGE

### 2.1. Telomeres.[155, 156]

The role of telomeres was first uncovered by Hermann Müller. He showed that a *Drosophila* chromosome without an end is unable to recover.[157] Barbara McClintock, in her work with unstable corn chromosomes, hypothesized the existence of normal structures that ensure chromosome unity.[158] It was discovered that telomeres fulfill this function. This fact was widely acknowledged in the late 1970s, early 1980s.[159] It was also shown that the chromosome segments closest to the telomeres are especially prone to change. This is evident in chromosome mixing which can be observed, for instance, in the formation of sperm cells.

Telomeres are structures situated at both ends of each chromosome. Besides their role in preserving chromosomal structure, other important features of telomeres were discovered. In the early 1970s, A. Olovnikov advanced a hypothesis that telomeres are crucial in cell division - they send division-controlling signals to the cell. With each division, a part of the telomeres breaks away and the cell ages, a process known as cellular senescence. Eventually, the chromosome runs out of telomeres, the cell can no longer divide, and it dies.[160]

This discovery seemed to open the door to understanding the process of aging and death. However, the situation turned out to be more complex. First, it was discovered that the results attained concerning telomere behavior did not apply to cancer or to sperm cells, the latter two are kind of immortal because telomeres are constantly replenished.[161] At the more advanced stages of the developmental-tumor, this waning of telomeres may terminate and they may even begin to recover. Telomere restoration is due to the activation of a cell factor or enzyme called telomerase. It contains the RNA structure which codes for respective new links of telomeres. In a normal adult cell telomerase is inactive.[162]

All of these characteristics of telomere behavior are not exclusively responsible, either directly (terminating cell division) or indirectly (the growth of telomeres in cancer cells), for an organism's death. As previously noted, biologists have recently uncovered two genes (ced-3, ced-4) in a drosophile cell whose joint action can result in a cell's death.[163] There might be a fascinating link between these genes and telomeres and telomerase!

However, my hypothesis regarding the internal somatic mechanism of change would lead one to think that telomerase activation observed in cancer and sperm cells pertains to all cells undergoing change. It is no surprise that cancer and sperm cells exhibit increased telomerase activity; these cells are most amenable to change. Cancer cell changes, almost by definition, and the changeability of the sperm, as I hope to show in the next chapter, is related to their role in the biological mechanisms aimed at development (rather than just survival or growth).

The aforementioned function of male and female gametes in the process of change and the difference in their respective volume of production are also related to the way these cells are formed (process). At the embryonic stages of development, ova are formed through mitosis of immature cells called oögonia. After a short while (in human beings by the third month), they turn into oöcytes which are the result of the meiosis process. At birth the ovaries house up to 400,000 *oöcytes*. As the organisms develops, these oöcytes merely mature. On the other hand, male gametes undergo several stages of mitosis, and by the time they reach puberty they turn into mature spermatazoons. Based on the data presented by James Crow, University of Wisconsin, by the time an organism is physiologically ready for mating, the male gametes have undergone 36 divisions, at the age of 20 - 200 divisions, and by 30 - 430 divisions, and at 45 - 770 divisions.[164] Normal human cells divide from 50 to 100 times.

It would seem that the necessary, but insufficient, condition for the life of an adult organism is the normal process of change, while a sufficient condition for death is either termination of change or pathological change.

The above statement may sound rather trivial - it is part of folk wisdom. Note, however, that widespread among businessmen is the notion that terminating a firm's development change is a sure way to bring about its demise.

Pursuing the idea that a cell undergoing normal transformation restores its chromosome telomeres to ensure continual reproduction, I believe that promoting normal change in an organism serves to increase its life span. This kind of approach differs from the one espoused by most scholars dealing with the problems of aging who advocate direct manipulation of telomerase.[4] Direct activation of telomerase is bound to disturb the balance of contingent physiological processes if the mechanism which turns it on under normal circumstances is ignored. I would also like to note that the idea of activating telomerase and increasing an organism's life span by supporting the mechanism of change bears in a very practical way, upon our attitude regarding cancer.

## 2.2. Possible Migration of Somatic Cells

The mechanism of somatic change, as described in the previous chapter, suggests that an innovator cell must coordinate its activities with cells of other organs.

This aim can be accomplished by dispatching cells containing new genetic information to other organs. One method to implement this kind of information exchange is by means of viruses that are, possibly, an intrinsic part of the information contained in the genome. This phenomenon is especially prominent in pathological cases (see the following Chapter).

On the other hand, this kind of limited information exchange might be inadequate, especially when the changes in the innovator cell are profound. Perhaps it takes total contact if the information to be transferred transcribes not only much of the genome, but also the cell's cytoplasmic structures.

Assuming that cell contact is required, there must be a mechanism of cell migration, i.e., separation of innovator cell from the host tissue and its penetration into other organs whose cells must undergo respective changes. Moreover, cell migration is probably non random, because the original host organ and the target organs are either morphologically linked or linked by the genetic sequence governing the development of a new organism.

At this point in time, this rather nebulous picture of normal cell migration is a fruit of my imagination. The prevailing opinion among biologists is that such a phenomenon does not take place. The study of the migration of cells in multicellular organisms was limited to such normal cells as embryonic, blood, lymphocyte cells and, in the pathological case, to the metastasizing cells. One special degenerative case of cell migration is the discharge of dead cells from the tissue. This case also involves a mechanism of cell cessation.

Generally speaking, the problem of cell migration remains largely a mystery [5], and it has not been proved that migration of normal tissue cells is impossible. In fact, in keeping with my approach of reconstructing the normal mechanism of somatic change based on its pathological manifestations, I would to say that cell migration could take place among innovator cells. Indeed, the phenomenon of metastasis suggests the existence of a mechanism that allows a cell to disengage from the host tissue and migrate to other organs. What prevents this mechanism from engaging when it comes to at least innovator-type cells?

One complicating factor in staging an experiment to test this hypothesis is determining the specific conditions under which the mechanism of cell migration becomes active. Moreover, this mechanism may affect only deviant cells, which would create an even more formidable task of distinguishing between innovator cells and the cancer cells.

## 2.3. Age-Specific Features of the Mechanism of Change

It appears that the nature of changes in cell structure differs with different stages of life.

During the embryonic stage of fetus development, the main concern is with the formation of an organism according to the inherited genetic code and the environment, including the womb, in which the fetus is developing.

During childhood, prior to puberty, the focus is on growth, basically within the framework of the inherited genetic information.

The process of cell change, especially a major change, belongs to the period following puberty, but prior to the start of the aging process. It is probably during this period that changes within the cells, as well as the interplay between the cells and the immune system including the transfer of new information to the germ cells, is best coordinated.

As the organism begins to age (perhaps excessive stress has a similar effect) break-downs in the mechanism of cell change, in the immune system and in the coordination between the two, become more aggravated. We cannot rule out that some negative consequences of aging are also due to the failing mechanism of cell change. In keeping with Dilman's theory of aging and major non-infectious diseases,[165] it is quite plausible that aging, like such ailments as high blood pressure, diabetes, cancer, etc., reflects an organism's inability to cope with the persistent flow of various nutrients needed at some point in time for the organism's growth. This inability may be due to the deteriorating mechanism of cell change.

Perhaps, in very old age, the mechanism of change ceases to function and cells that fail to change simply die out.

## 3. QUASI-HYPOTHESES STEMMING FROM THE MATERIAL PRESENTED IN THIS CHAPTER

1. The genetic system of a cell possesses a hierarchically organized internal mechanism of change. The program that codes the organism's development is denoted as the zero level program; the program that changes it represents the first level program; and the program that changes the first level program is termed the second level program.

2. Taking into account the intricate nature of the genetic system and the resources required to sustain it (see, for example, the C-value paradox in Chapter 3) an organism's development raises the problem of resource allocation between creating and nourishing a dynamic genetic system, on the one hand, and, on the other hand, faster cell growth (in conjunction with greater quantity) governed by a routine genetic program.

3. It is quite feasible that the genetic system incorporates the tunnel process. In other words, change is initiated both at the end, meaning it is induced by the environment that can directly affect the zero-level

program via chemicals, radiation, etc., as well as at the beginning, i.e., the internal changes in the second or first level programs.

4. With changes initiated at the beginning, there ought to be structures in the genome that minimize those genetic combinations that lead nowhere or to a dead end.

5. While the somatic mechanism of change plays a secondary role, perhaps it continues to fulfill the following functions: a) complements the germ mechanism in situations where relatively minor and slow changes are required by the organism, b) fulfills specific functions not covered by the germ mechanism, c) acts as a back-up ensuring that such crucial evolutionary function as change is not neglected, and d) finally, it may be an anachronism.

6. Minor changes unfold primarily from the end, while major changes originate at the beginning.

7. In trying to reconstruct the normal mechanism of somatic change through its pathological manifestation, in this case metastasizing, one could hypothesize the migration of normal somatic cells; perhaps some emigrant cells even return to the mother organ. Moreover, cell migration is not chaotic in the sense that cells from one organ get attached to those organs that are logically connected by the process of change. The fact that even sporadic migration of normal somatic cells has not been documented does not mean that such a phenomenon does not exist. A number of biologists have confirmed that somatic cell migration does yield to experimental verification.

8. Presumably, if the changes in the cell are of innovative nature, the repair mechanism could be turned off – it either ignores innovations or activates genes that block the repair mechanism.

9. Since activation of the telomerase pertains to all cells undergoing change, natural death or fading away of an organism is possibly due to the termination of the process of change. A number of scientists have entertained the idea of extending an organism's life span through direct manipulation of the telomerase. It seems that direct activation of telomerase, without touching the mechanism which triggers it under normal conditions, may derail other related physiological processes.

## NOTES TO CHAPTER 6

[1]. Shapiro classifies the repair mechanism as a mechanism of change.[166] In seems to me that it should be assigned to damage-type changes (break downs, mistakes).

[2]. There can arise other situations when a cell sacrifices itself in the name of the overall development. In this context, apoptosis may be regarded as a manifestation of a cell's altruism. The phenomenon of apoptosis (similarly to self-sacrifice in the social realm) is limited to a small number of cells. Just a note in passing: there is not a single instance in the Torah describing self-sacrifice, for whatever noble purpose.

[3]. Presumably biochemical processes need both catalysts and inhibitors. Generally speaking, both accelerators and brakes are needed to keep any strong fluctuations under control (as in the whip and the reins used to control a horse). I have applied this generalization to economics, suggesting that faster growth of the economy requires both income growth - a catalyst, as well as price increase - an inhibitor (within reasonable limits, of course). More in my article. [167]

[4]. "...if manipulation of cellular life span were technically possible through regulation of telomerase, it might be possible to decrease morbidity or increase life span of the organism." [168]

[5] "How cells migrate has been the subject of much scrutiny and model building for many years, but is still not understood." [169]

# CHAPTER 7

# SOME FEATURES OF THE GERMATIC MECHANISM OF CHANGE: THE ORIGINS OF SEX DIFFERENTIATION

To reiterate, the leitmotif of the present book is the mechanism of change incorporated in the somatic cells. Allow me to digress and pick up a parallel motif reflecting the external mechanisms of change implemented via the germ cells. I believe the digression is justified for I shall bridge this minor theme with the leitmotif in the later part of the book.

The subject I want to touch upon is one particular mechanism of change based on the germ cells - the mechanism of sexual reproduction.

There are many ways to classify methods of reproduction. Interesting in this connection is the typology proposed by A. Kondrashov. His approach was based on population genetics which he regards as a crucial element of evolution. His classification distinguishes two methods of reproduction: amphimixis and apomixis:

> "amphimixis - that is, a life cycle with alternating syngamy and meiosis - advantageous over apomixis - or production of offspring from single mitotically derived cells. Amphimixis is used instead of the potentially confusing term "sexual reproduction" because sexes are exogamous classes of gametes. In some cases of "sexual reproduction", any pair of gametes can form a zygote (e.g. homothally in fungi), and there are no sexes. Similarly, apomixis is used instead of "asexual reproduction" because the latter also sometimes includes vegetative reproduction when a progeny appears from many cells."[170]

The proposed taxonomy of crossing is based on their chronological evolution of reproductive methods. My initial criteria generates two categories - somatic (based on fragmentation) and germatic (based on specialized germ cell). Germatic reproduction is further classified based on the number of cells partaking in the reproductive process (one cell (*spore*) or cell crossing). The cross-over category is further subdivided based on the following criteria: are the cells homogeneous or heterogeneous (sexual) in terms of function. Finally, heterogeneous cells split into two classes depending on whether they are contained in a single organism (hermaphrodite) or carried by different organisms.

Presumably at the early stages of evolution simple multicell organisms lacked germ cells and reproduced by means of fragmentation. The process of change in such organisms was rather cumbersome since change in some cell(s) had to be harmonized with changes in other cells of the organism, whether these changes were induced internally or by environmental conditions including hybridization.

It seems the evolution "realized" that change can be implemented more effectively via the germ cells if only because the genetic material which defines the development of a new organism is collected in one compact space. Over the course of evolution, reproduction via germs underwent fundamental shifts. Evolution initially devised germs and then separated the hosts of these cells, i.e. creating genders. The process of evolution has not ended and new forms of crossing may emerge in the future.

While there is no general agreement among biologists as to the origins of the sexes one major benefit of sexual differentiation advanced by biologists is the increase in diversity of genetic combinations generated by crossing. Thus, *compatibility* of the crossing organisms represents a necessary condition for the emergence of the sexes.

In my opinion, however, the above condition is not sufficient for the emergence of sexes. I have tried to elaborate a general approach to sexual differentiation, not necessarily limited to two sexes, and to show that distinct functional attributes of crossing organisms represent another necessary condition in the definition of the sexes.

It should also be noted that the crossing mechanism is geared toward greater diversity, or at least toward preserving diversity, and may well come in conflict with population growth. This may occur, for example, when the number of sexes that partake in crossing exceeds or equals the number of offspring.[171]

Indeed, for population expansion purposes reproduction, when germ-cells are present, parthenogenesis would be most appropriate in terms of the above objective, and in case germ-cell fusion does take place, all the constituent genetic components should belong to one host organism (hermaphrodism). No matter how virile a given organism is, meaning whatever its capacity to bring new creatures into the world through a single act of crossing, all other conditions being equal, the method of crossing sexes produces less offspring than if the act was carried out by a single organism. Reproduction via a single organism produces at least one new offspring and the total population will be two, i.e., the proportion between the total population and the originators of offspring is 2:1. Now, even if only two organisms cross to produce a new one, then a progeny of one will generate a total of only three organisms, i.e., the proportion between the total population and the originators of offspring is 3:2.

The arithmetic simply underscores the fact that crossing via sexes is really geared toward the creation of diversity. Naturally, diversity itself may prove conducive to survival and growth if only because it gives rise to creatures that might be better adapted to changing conditions.

## 1. SEXUAL TYPES AND HOW THEY EVOLVED OVER THE COURSE OF EVOLUTION

### 1.1. Definitions

Greater diversity of life forms, being a necessary condition of biological evolution, can be attained via changes within a given organism as well as via mutation and recombination of genes. The passing of changes in a given organism to the progeny is accomplished by means of fragmentation or germ cells (for single-cell organisms the notion of a germ cell and an organism is one and the same). In turn, each germ cell is ether a spore, or a carrier of gametes, complete hereditary information. Organisms' recombination under fragmentation (division) is carried out by means of hybridization. With sex cells present this function is fulfilled by crossing (and sometimes by hybridization.)

Whatever definition of sex we pick, a necessary condition for sexual differentiation is the compatibility of the crossing cells. At this juncture it does not matter whether these cells fulfill different functions, whether crossing takes place in one or more organisms, or whether the offspring is fertile or sterile.

I would like to elaborate on the above statement and examine the crossing mechanism in some of the simplest organisms (which replicate via fragmentation) that lack functional differentiation but whose structure incorporates certain physiological variations (examples of such organisms are given below). In fact, these simple organisms must be compatible and not every member of the species can cross with any other member. The actual compatibility may hinge upon the effectiveness of crossing of physiologically different organisms even if we assume that they can cross with each other, at least in theory. It is quite plausible that what we observe here is a very characteristic phenomenon in biological systems, namely, prevention of less effective solutions.

To sum up, we want to isolate from the manifold of organisms that change by means of crossing a subgroup which shall be termed compatibles.

Another *necessary* condition for sexual differentiation (for now omitting the explanation) is for the cells involved in recombination to be functionally distinct. Thus, a cell's affiliation with a particular sex would depend on the specific qualities attached to the "compatibles", i.e., the functions assigned to each sex other than just the capacity to cross. Therefore, functionally different germ cells are compatibles but not vice versa.

One more important point is that the presence of different germ cells is not sufficient for the emergence of the sexes since germ cells may be produced by a single organism. By definition sex refers to organisms that carry a single type of germ cell (gametes).

## 1.2. Examples

The significance of this definition of sex may be illustrated with other relevant combinations between the germs and their hosts. The combinations incorporate the following parameters: the number of organisms that produce different germ cells, the number of organisms that take part in producing offspring immediately after the germ cell is produced, and the number of organisms that partake in crossing. The table below presents different types generated by the various combinations of these characteristics.

TABLE 7.1. Germ cells and their hosts.

| The number of producers of different germ cells | The number of producers of new organisms immediately after germs are produced | The number of organisms partaking in the crossing act | | |
|---|---|---|---|---|
| | | One | | Several |
| | | Parallel | Sequentially | |
| One | One | Full hermaphrodite | X | Symbion pandora |
| | Several | X | Pseudo-intersex | Hermaphrodite of herms type |
| Several | Several | X | X | Sexes |

Let us examine the various types of organisms presented in the table.

### 1.3. Hermaphrodite

A "hermaphrodite" is an organism that contains different gametes. The term "full hermaphrodite" shall be reserved for self-fertilizing organism that produces progeny (not to be confused with parthenogenesis when there is no self-fertilization; simply one female sex is sufficient to produce an offspring). This kind of full hermaphrodism is characteristic of relatively simple forms of life such as snails.

This suggests that even full hermaphrodites are merely a stage in the development of sexes.

Interesting in this respect is the classification of hermaphrodite types proposed by Anne Fausto-Sterling.[172] Between what I termed "full hermaphrodite" and sexes there are at least three other groups, all incapable of self-fertilization, united in the medical literature under the term "*intersex*". Fausto-Sterling includes in the first group called *herms* that have one testis and one ovary. She purports that in principle herms could perform as both fathers and mothers but their system of ducts and tubes prevents the fusion of sperm and egg. The second group is comprised of male pseudohermaprodites called *merms* that have testes and incomplete female genitalia but no ovaries. The third group of female pseudohermaprodites called *ferms* possess ovaries and some features of male genitalia but no testes.

Together with the male and female sex all these hermaphrodites comprise, according to Fausto-Sterling five sexes.

Fausto-Sterling's work is interesting in that she focuses on the diversity and special features of hermaphrodites, an area which is rather obscure. However, aforementioned arguments supporting my definition of sex conflict with her classification that matches various hermaphrodites to different sexes.

Our examination of hermaphrodites provides a new insight into the origins of the sexes, including male and female sexes. Biological literature on the subject (as does mythology) frequently raises the question of primacy, i.e., which sex originated from which. This issue is still unresolved with contrasting views being expressed even when it comes to the origins of male and female sexes.

> "In the organizational concept the female is the default sex and the male the organized sex, imposed on the female by the action of hormones. In my alternative scenario, the female is the ancestral sex and the male the derived sex."[173]

Our discussion of evolution suggests that sexes emerged as a result of a single germ cell splitting up into two functionally distinct cells both housed in the same self-fertilizing organism (full hermaphrodite); subsequently hermaphrodites gave rise to sexes, meaning functionally specialized hosts of specialized germ cells. Attesting to the plausibility of this course of development of the reproductive system are the vestiges of hermaphroditic phenomenon still present in sexually differentiated organisms.

### 1.4. "Sex-convertible Organisms"

The spectrum between full hermaphrodite and sexes contains so called *sex-convertible organisms*. In terms of sexual differentiation, organisms that belong to these intermediate forms have no special functional features

## THE ORIGINS OF SEX

although theoretically they that can fulfill the functions of either sex. Such sexual transformation is observed among some species of fish that can alternate between male and female sexes depending on the circumstances.

Sex-convertible organisms, as well as ordinary sexes, are capable of fulfilling the functions of one of the sexes in a single act of crossing. At the same time, they differ from sexes in that the latter are specialized while sex-convertible organisms are universal. By the same token, sex-convertible organisms, as well as hermaphrodites, are capable of fulfilling different functions in the act of crossing. But, while hermaphrodites can do so, at least in principle, in parallel, the sex-convertible organisms can do so only in sequence, i.e., after undergoing the necessary transformation.

### 1.5. Symbion pandora

This organism attaches itself to lobster's lips and alters its mode of development as it evolves. At the asexual stage, *Symbion pandora* is a larva which produces both male and female cells similarly to hermaphrodites. The larva evolves into a midget male organism replete with spermatozoids and housing a developing female organism. Subsequently, the female leaves its host, and, as the female dies, it discharges a fertilized cell which develops into a new larva. The old larva does not die after the sexes are formed but continues to develop until it reaches a stage at which it discharges males. Perhaps this intricate fusion of sexual and asexual stages of reproduction suggested to D. Ackerman, the author of an article about *Symbion pandora*, the term "trisexual, it will try anything".[174]

It seems that this mode of reproduction represents a peculiar combination of sexual and asexual processes rather than true trisexual reproduction. This class of organisms is very interesting and could be termed *chimera*.

The list of possible combinations between the germ cells and their hosts that partake in crossing has not been exhausted. But my key point is to state a sufficent condition for the definition of developed sexual types, namely the emergence of functionally specialized germ cells carried by specialized hosts.

As revealed in the next section many biologists reduce the problem of sexual types to genetic recombination.

## 2. THEORIES OF SEXUAL TYPES THAT FOCUS ON THE IDEA THAT CROSSOVER TAKES PLACE VIA VARIATIONS OF GERM CELLS

The functional diversity of germ cells is usually ignored, meaning most theories fail to recognize that the capacity for cross-over is just one necessary condition for the definition of sex and that other criteria are no less relevant. "Sex is the process whereby a cell containing a new combination of genes is produced from two genetically different parent cells."[175] (p.87)

Thus, whatever definition of sex is picked by biologists, from the functional point of view, the origin of different sexes is identified with greater genetic diversity within the species with little or no importance assigned to the functional specifics of the organisms involved in crossing. Still, it is not denied, in fact, it is presumed that the ultimate aim of diversity is quantitative growth of the species.[176]

This approach to sexual types is well presented in a famous work by Edward Wilson Sociobiology. Interestingly, Wilson preludes his discussion of the origins of sexes with what seems at first glance a very paradoxical remark. "Sex is an antisocial force in evolution"[177] (p.314).

He goes on to say that

> "Perfect societies, if we can be so bold as to define them as societies that lack conflict and possess the highest degrees of altruism and coordination, are most likely to evolve where all of the members are genetically identical. When sexual reproduction is introduced, members of the group become genetically dissimilar.
> ... The inevitable result is a conflict of interest.
> ... It has always been accepted by biologists that the advantage of sexual reproduction lies in much greater speed with which new genotypes are assembled." (p.314-315).

An example of the notion of recombination being confused with different sexual types is the recently discovered process of reproduction in Fungus.[178] A more complex case is reproduction in *Paramecium*. It was observed that *Paramecium* engage in cross-over which ensures greater genetic diversity. It has also been established that under relatively stable conditions *Paramecium* reproduce by division. Under changing conditions they begin to mate thereby merging and exchanging genetic information

# THE ORIGINS OF SEX

(primarily through nuclear fusion; afterwards the "compound" *Paramecium* stretches out and forms a dumb-bell; finally it splits and forms two *Paramecium* with new genetic traits which then reproduce by division).

However, Paramecium do exhibit selectivity - not every organism mingles with any other. I quote T. Sonneborn, a pioneering specialist in the field of *Paramecium* reproduction.

> "We have studied the inheritance of three kinds of traits in *Paramecium*. In all three respects the two mates of a pair do, in fact, produce unlike cultures. One of these traits is sex, or mating type. Although the two individuals that mate are functionally hermaphroditic, they differ physiologically: mating can occur only between physiologically diverse cells, that is between different mating types."[179]

In stressing the role of the sexes in preserving diversity, the theory of mating types advanced by H.Bernstein, et al,[180] is actually a particular case of the general approach rooted in genetic diversity.

Their theory states that the female sex is actually sufficient for reproduction. However, there is a multitude of external factors such as radiation, chemicals, etc. that exert a damaging effect on the ovum. The function of doubling the number of chromosomes through fusion with the male germ cell is an effective restoration of the damaged genetic structure of the ovum.

However, the need to repair damage through genome doubling does not warrant the emergence of such complex interaction between different mating types. First of all, the cell itself possesses a mechanism of self-repair. Even if this mechanism is not adequate, chromosome doubling, as means of repair of the genetic structure of the ovum, could be achieved in a simpler manner through the fusion of germ cells belonging to similar organisms, not to mention the fact that the germ cells of organisms having asexual reproduction, such as certain species of lizards, have significantly more chromosomes than developed mammals. Moreover, all the chromosomes in the cell come in pairs, so for repair purposes, asexual reproduction would perform just as well.

The functional aspect of mating types is linked with the structural aspect, which is predicated on structurally differentiated organisms. Of course, structural differences range from minor genetic variations to major differences that define the function of the mating organisms. In speaking of the sexes I want to emphasize the differences in the reproductive system of the organisms directly partaking in mating (unlike differentiation based on the organism's function in a colony - workers and soldiers among insects).

From the process-oriented point of view, sexual differentiation depends on the number of participants involved in cross-over. In nature we observe mainly one particular case of pair-wise crossing (mating).

> "Why are there usually just two sexes? The answer seems to be that two are enough to generate the maximum potential genetic recombination, because virtually every healthy individual is assured of mating with a member of another (that is, the "opposite") sex."[181] (p. 316).

Whith other conditions being equal, mating may indeed produce the greatest number of new genetic combinations given the size of the population. However, taking other factors also conducive to genetic diversity into account, it may turn out that the act of crossing could well involve more than two organisms. I shall come back to this question in a later part of the paper. At this point I just wish to note that the general notion of sex as a compatible creature with different genetic make-up would point to multi-sexual reproduction among flu viruses. These viruses were observed in genetic recombination that involved two or more viruses.[182]

Let us now look at the category of sexual types in terms of genesis. Interesting in this regard is the theory advanced by two Canadian scholars, Michael Rose and Donald Hickey. They claim that the emergence of sexes did not provide any evolutionary advantage. In fact, they portray sexes as a mechanism that promotes the spread of parasite segments of the DNA; these DNA sequences found in many cells are similar to viruses and are commonly found in many types of cells."In this context, males can be seen as parasitic DNA made manifest at the organismal level."[183]

The authors further contend that while sexes originated as parasites they evolved into biologically useful entities.

In the light of my concept of the tunnel process and sexual types (see section 3) the above theory gains a new dimension. Perhaps, the authors' contention that sexes evolved from the beginning rather than from the end is correct. In that case the initial stages of innovation (here, the sexes) ought to be carefully tested, meaning the novelty which at first seems useless or parasitic may, in fact, turn out to be beneficial. If we make the opposite assumption regarding random mutations, i.e., if we rule out that there are mechanisms in nature that form biological potential, then nature's verdict on these pragmatically obscure states will, at least initially, be negative. Of course, positive evaluation of biological potential is a very complex task as revealed in our discussion of the tunnel process.

Now I want to examine in detail three theories of sexual types, the last of which is my own.

## 3. THEORIES OF SEXES THAT FOCUS ON THE IDEA THAT CROSSOVER TAKES PLACE VIA FUNCTIONALLY DISTINCT GERM CELLS

### 3.1. Laurence Hurst's and William Hamilton's Theory of Sexual Types

One contemporary theory on the nature of sex links the emergence of sexual types with conflict prevention in the zygote between different gamete parts that contain genetic information.[184] This concept incorporates the idea that structural elements bearing genetic information are not limited to the cell's nuclei but are contained in organelles (e.g., mitochondria) and other cytoplasmic structures.

Just as any other theory the above theory has undergone its own evolution, most comprehensively expounded in the works of Lawrence Hurst and William Hamilton.[185-188]

Essentially this theory states that sexes differ based on the extent to which an organism directly partaking in crossing via its germ cells is able to shed its cytoplasmic genetic structures including organelles. According to the authors, the importance of organelle shedding as a sexual characteristic stems from the fact that the zygote is not some idyllic system but an active and inner-conflicting one. This turbulence is a result of different parts of the germ cells competing with each other when they fuse because each cell pursues its own interest. One special threat to zygote development is posed by the competition from the organelles which, unlike the chromosomes, are not integrated with the zygote.

The outlined theory purports that "sexes" appear when the zygote is formed through a complete fusion of germ cells belonging to different organisms. For example, this process takes place in common green algae *Chlamydomonas*.

On the other hand, according to the above theory, sexes are non-existent when the interaction between germ cells is limited to the exchange of nuclei. According to the authors, sexual differentiation in many species is predicated upon one sex sacrificing its organelles in order to form a zygote. The most widespread phenomenon is the hierarchy of two sexes when one of the two partners, usually the male sex, gives up its organelles.

The authors contend that their criteria allows for multi-sexual reproduction. Certain species of slime mold, e.g., *Physarum polycephalum,* could have as many as 13 sexes. This approach to sexes is based on a multi-level hierarchy of organisms. Under fusion, the organelles of the higher organism are passed on; the lower ranking organism gives up its organelles. As I see this case, what we are dealing with is not 13 different sexes but a more selective kind of crossing among compatibles. However, from the evolutionary perspective this intricate mode of crossing is not expedient. Employing his definition of sexes, Hurst notes :

"For any particular sex the cytoplasmic genes sometimes will be inherited, and sometimes won't be inherited depending on who you mate with, so it's got an inherent vulnerability to cheats; what happens if one mutant set of mitochondria refuses to shut down?"

The authors thus conclude that two sexes are optimal.

In lieu of the recent discoveries pertaining to gamete recombination Hurst had to revamp his theory on the origins of sexual types.[189] It was discovered that complete fusion of male and female gametes was observed not only in mussels (*Mytilis*) but also in mice and *Drosophila*, although in a less developed form. In fact male and female offspring of mussels are endowed with different mixture of father's and mother's mitochondria. For example, daughters usually inherited just the mother's mitochondria while the sons primarily inherited a combination of mother's and father's mitochondria.

As is the fate of many theories, subsequent discoveries not so much invalidate their predecessor theories as limit the scope of their application to a more narrow range than was at first suggested by the author. Hurst has said that his original theory of sexual types was no exception.

## 3.2. V. Geodakian's Theory of Sexual Types

There are other approaches to the definition of the sexes and their role in evolution. The theory of two sexes proposed by Geodakian emphasizes the benefits of mating which is claimed to produce an optimal combination between population size and genetic diversity within the population.[190] The key to Geodakian's theory is the general cybernetic notion that organized systems exhibit:

> "Separation between the task of preserving (conservative body) aimed at keeping things as they are and the task of changing (operative body).
> ...the implications for biological systems are the following: females in the population exhibit stronger tendency for hereditary continuity, while males - for change.
> ...to use cybernetic terminology, females represent the virtual memory of the species and the males manifest the operational (temporary) memory of the species."

Within the outlined framework Geodakian attempts to explain certain facts related to the number of males in a population. For example, it was observed that under favorable conditions very simple creatures having sexual reproduction such as water fleas and aphids exhibit a strong tendency for parthenogenesis. At the same time, in a more turbulent (changing) environment some females begin to produce males who then fertilize the females. Moreover, the observed difference in male and female roles sheds light on the accelerated turnover of males in the population, i.e., their higher rate of birth and death under changing conditions.

Geodakian notes that the number of males in a population may increase as a result of some females turning male, a hypothesis supported by some recent research in the West. It was observed that fresh water hermaphrodite African snails alter their pattern of reproduction depending on the density of parasite organisms in the water. If the water is free of parasites the snails fertilize themselves. However during intense infestation some snails grow a male organ and engage in sexual intercourse[191].

Unlike Hurst and Hamilton, Geodakian's theory of the sexes is oblivious to cell structure. In other words, the cell function is not linked in any way to cell structure. This omission was the result of the following considerations:

"...differentiation into two sexes ensures the production of two kinds of gametes or germ cells: tiny and mobile spermatozoons whose task is to come in contact with a relatively large but stationary ovum that provides nutrition for the future embryo."

Geodakian's stance on the subject basically falls in line with another theory that links the origins of different sexes to differences in gamete type: one type is relatively simple therefore behaving as attractor gametes; the other type is complex acting as attracted gametes.[192] I agree with Geodakian that sex differentiation is not rooted in these differences. Rather the fact that these cells exhibit different levels of activity is a consequence of their functional role, a topic discussed below.[1]

It seems that the theory of sexual types developed by Western scholars is similar to the one proposed by Geodakian: it basically emphasizes the need to eliminate the impact of parasitic entities.[193] However the authors of all these theories based on the emergence of the sexes as a response to changes in the environment fail to mention Geodakian's name.

Both theories of sexual types (Geodakian's and Hamilton's/Hurst's) are impressive if only because each theory distills certain important empirical facts and suggests a plausible explanation. By emphasizing different aspects of sexual reproduction the two theories seem to complement each other.

## 4. MY THEORY OF SEXUAL TYPES

### 4.1. Multisexual reproduction

My approach to the concept of sex is based on the role of individual participants of the crossing act in fulfilling the function of change (this is different from classification based on unchangeable specific function inculcated in the organisms, e.g., workers and soldiers among insects).

The outlined approach to sexual types allows for more than two organisms to partake in cross-over. Most theories of sexual types, as it was mentioned above, postulate only two sexes. The reason I question this particular assumption is because its revision gives us a broader view of the problem of mating and the sexes.[2]

Other conditions being equal, mating is optimal for implementing crossing between functionally distinct organisms because it minimizes the number of such organisms. However, other mechanisms of crossing involving more than two organisms can be construed.

As I have noted above, under mating population size is sacrificed (relatively to hermaphrodism) for the sake of greater genetic diversity. Similarly, with three or more organisms required to produce a new one, population size and even the number of genetic combinations may well decline, but the "quality" of the combinations actually produced may ultimately prove more conducive to the development of the species.

On May 3, 1984 I was invited to give a talk at a "crazy ideas" seminar held at the Benjamin Franklin Research Institute in Philadelphia. The subject of my lecture was "Multi-sexual reproduction," meaning reproduction involving more than two organisms.[3] I chose to use an analogy with political systems where the birth of a new social institution calls for a deep-wrought separation of powers. Using the above analogy, we could imagine at least three sexes.

The first sex, counterpart of the legislative branch, would elaborate programs of strategic development with long-term significance. By analogy with the executive branch, the second sex would elaborate tactical programs within the framework established by the first sex. (Presumably, adjustments to current conditions – "operative management" – is carried out by the organism using the means at its disposal, such as reserves, organs for adjusting to temperture fluctuations, etc.) Finally the third sex, the "judicial branch", would confirm that the programs followed by the other two sexes are in accord with the fundamental programs of development, thus preventing the birth of organisms violating basic precepts of development. The idea of prevention in biological development is quite plausible: it is unlikely that "quality control" over new structures is performed exclusively by hindsight, i.e., through natural selection.

So, my approach to the definition and classification of the sexes implies that the organisms directly partaking in crossing must play different roles, analogous to different functions fulfilled by the governing bodies. In this respect my approach resembles that of Geodakian. In fact, the role he ascribes to the male sex is essentially the same as the second sex,"executive power", in my scheme. However, the function of the first sex,"legislative", which might correspond to the female sex is vastly different from the role ascribed to the female sex by Geodakian. He deems female the conservative sex - it preserves what has been attained. In my opinion, the female sex takes on development geared toward profound long-term changes (I shall elaborate on this point later on).

As far as the third sex is concerned, I am not familiar with its equivalent in nature. However, the existence of the third sex with its respective functions is quite plausible. If one was to seek out the third sex in nature one should probably do so among organisms where the population density is rather high making it easier to meet representatives of all the sexes needed for reproduction. Even if not found in nature we can set up computer simulation of the evolutionary process with tri-sexual reproduction and see what kind of results are obtained.

## 4.2. The Tunnel Process and the Sexes

To reiterate the point made in our previous discussion of the R&D process in economics, the process of innovation represents a multi-stage process which can start at the final stage (end of the tunnel) that reflects the immediate concerns of the environment, or it can start at the initial stage that echoes the immediate concerns of internally-driven development of basic science. Still owing to the mutual feedback among its various stages, R&D represents a single unified process .

I have also noted that an analogous tunnel process phenomenon is observed in biological systems. This approach might help explain the splitting of specialized germ cells which occurred over the course of evolution as gametes and sexes appeared, first and foremost among the developed mammals. One working assumption is that under mating the male sex voices change associated with the final stage of the process of development, i.e., it reflects the demands of the environment. The female sex, on the other hand, is the vehicle of change characteristic of the initial stage, i.e., it implements profound internal shifts in the structure of an organism. The above scheme does not preclude the male sex having any connection with the initial stage or vice versa for the female sex.

The following puzzling question served as the impetus for these speculations: "Why is it that in developed species the testicles are outside the body and the ovaries hidden deep inside the body?" The first part of the question is well explored. The common explanation is that sperm production takes place at temperatures below the so called normal body temperature. In my opinion, this explanation can be challenged. For one thing, being outside the body makes testicles, a rather delicate organ so vital for procreation, rather vulnerable.[4] As far as the temperature factor, an organism could have easily created a cooler niche inside itself to accommodate sperm production. We know that various parts of the body exhibit a wide range of temperature and in some species of mammals, e.g., elephants, whales, dolphins, the testicles are hidden inside.

It seems to me, that this kind of fragile construction of the male body is justified if, by being outside, the testicles would be more susceptible (responsive) to changes in the environment and especially such a global parameter as radiation. It has been determined that changes in earth's radiation level (sometimes as a result of sun spots) are linked with certain biological phenomenon such as migration of lemmings, crop yields, etc. The fact that female ovaries are well hidden under the skin protects them from some kinds of current external factors allowing them to focus on the other mechanism of change.

All these speculative deliberations can be reformulated into experimentally verifiable hypotheses. For instance, the following experiment could be set up: a small dose of radiation could be applied to the testicles to see whether that affects the sperm. The females could be subjected to a similar small dose of radiation and then the ovaries and the ova checked for the aftereffects. If this kind of radiation exposure produces different results in the reproductive organs of male and female animals then my hypothesis regarding higher absorbency of environmental factors by the male gains credibility.[5]

Still, there is other evidence supporting the notion of functional differentiation of the sexes. I want to emphasize that the mutation rate is much higher in the human male germ line than in the female germ line [6] and, to bolster the claim, there are profound dissimilarities in the reproductive systems of males and females so strongly expressed in humans. With each ejaculation the human male releases up to 300 million sperm, while ovaries of a new born female contains about 400,000 ova and no new ones are produced.[194]

General principles of evolution point to at least three factors that allow species to develop: 1) capacity for quick reproduction; 2) organisms' ability to rapidly adapt its structure to the new conditions; 3) active individual performance.

There is a strong correlation between these factors and an organism's complexity. The more complex an organism, the more pronounced is the third factor and the less critical are the first two.

Indeed, for viruses, microbes, bacteria, etc. a typical organism is relatively simple. Owing to its simplicity these organisms are able reproduce at a very fast pace and some are able to alter their structure within a short period of time (viruses can do this literally within years). On the other hand, each individual organism may easily perish in the course of its life cycle. As organisms become more complex the reproduction rate wanes and the time required for structures to change increases. Nevertheless, these parameters are respectively considerably greater/less in insects, for instance, than they are in mammals. Each mammal is relatively complex. Its reproduction cycle is rather long and changes in its structure are very slow. At the same time, each individual creature is extremely adept and endowed with a variety of means to fulfill its function.[195]

Comparing sperm and ova one observes certain correlation between their complexity, factors supporting their functional role, and the coordination between their function and other aspects of their system's performance. From the functional point of view, the primary role of the spermatozoid is to convey changes in the environment to the succeeding generations by doubling the number of chromosomes, triggering the development of the fertilized cell, etc. If the sperm's structure was a complex one ( i.e., instead if millions of sperm there was one very complex one) it could still function as far as interaction with the female members of the species. However, the creation of such a complex sperm would not agree with its alleged function of rapid change. The fact that the sperm is structurally similar to viruses is precisely the feature which allows it to change rapidly.[7] The process of sperm production entails multiple divisions of the sperm. This kind of behavior promotes diversity since many changes take place during cell division. Perhaps, the genesis of sperm formation over the course of evolution can be traced by observing the process of sexual differentiation as the embryo develops.

Assuming the primary function of the ovum is to support major changes in the organism, passing on the genes to the progeny is a necessary condition for this function to be fulfilled. The cell's structure would have to be complex to fulfill this function.

Quite possibly, internal genetic changes experienced by the ovum as it forms (process dimension) occur at the initial stages when the cell is not yet mature (oögonia). The cell undergoes multiple divisions; external changes (but still within the organism) take place at a stage when the already indivisible ovum begins to mature (oöcytes). As far as the genesis facet of cell dynamics, prior to the emergence of different sexes the germ cell absorbed both the environmental factors as well as the internal changes. Perhaps, over the course of evolution the task of responding to current external factors was shed and assumed by the specialized cells of the male sex. Nevertheless, the ovum has kept all of its former functions. Parthenogenesis attests to the idea that unlike the sperm the ovum is, in principle, sufficient to ensure reproduction. Vestiges of parthenogenesis have recently been discovered even in humans[196, 197]

All these speculations regarding the germ cells, assuming mutations can occur at different stages of their development, is corroborated by J. Haldane's hypothesis that the ovum and the sperm of different species exhibit very different mutation patterns.[198] The mutation frequency pattern of male and female gametes of different species is the following: two in rodents, six in primates, and ten in humans.[199]

What is the significance of male gamete mutability, especially among the more developed animals?[8] A number of biologists have offered the following explanation for the reasonably high rate of mutation of the male gamete: this process makes up for the epistatic selection (selection where one gene acts to suppress another gene) which reduces diversity as well as the formation of harmful mutations. However, since excessive mutation rates may generate too many new harmful combinations, the task is to find some optimal number of mutations that will serve to promote the evolutionary process.[200] An interesting model to simulate optimal mutation frequency ratio between males and females was undertaken by R. Redfield[201] The results obtained are not conclusive since the model could have failed to take certain factors important to the evolutionary process into account. A large percentage of male gamete mutations may turn out to be harmless or useless because variation is limited to the so called "selfish" genes whose role is still obscure.

## 4.3. Characteristics of Sexual Types and Information Transfer Between the Somatic and Germ Cells

The above discussion of the relation between the tunnel process and sexes has lead me to another topic, the transfer of information from somatic to germ cells.

Suppose the migration of somatic cells in multicellular organisms does take place. In the extreme case a changed somatic cell might even try to invade the germ cells in order to pass on changes to the progeny. This whole notion is rejected by the great majority of biologists who hold that evolutionary changes are due to external factors affecting DNA of germs. Of course, in view of the grave danger of introducing untested changes in the established genetic program there must be strong barriers in the way of somatic cells penetrating germ cells. Perhaps, these barriers make it difficult to observe somatic cell penetration even if the phenomenon does take place. In principle, it could take place in a physiologically modified form through viruses going from innovative somatic cells into germs. It is fascinating to find out how viruses manage to accomplish this task.[202, 203]

Direct penetration by innovator somatic cells is more complex. The process might turn out to be quite different in males and females. In the light of such phenomenon as parthenogenesis it would be reasonable to suppose that change unfolds primarily though the female medium. My assumption that females impart major beginning-induced changes and males convey changes from the end seems to imply that in males positive change unfolds primarily through the germ cells. Perhaps, in the female sex the archaic somatic mechanism of change has been preserved to a greater extent than in the males: profound changes originating at the beginning might require more testing of the various components (at least at the informational level) and their subsequent integration - something the somatic mechanism of change accomplishes more thoroughly.

These speculations regarding the roles of the somatic and germatic mechanism of change agree with the following facts. There exist mysterious biological barriers in the way of foreign cells invading the scrotum but no such barriers exist for ovaries. As a result, secondary cancer of male testicles is non-existent (and primary cancer is rather rare) while ovaries (and the female reproductive system in general) are a favorite spot for secondary as well as primary cancer. It goes without saying that these facts establish a certain correlation rather than a proof of my supposition that innovator somatic cells should have an easier time penetrating the female reproductive system.

## 5. CONCLUSIONS

1. The development of the germ cells, prior to sexually differentiated cells, and the germatic mechanism of change that succeeded the somatic one on the evolutionary scale was dictated by the advantages of implementing the process of change within one type of cells - *in one place* where it is much easier and faster to coordinate all the changes.

2. Sex can be defined based on the role of specialized germ cells directly partaking in the act of crossing by means of specialized hosts.

3. A question that frequently comes up is which sex originated from which. It seems that sexes emerged as a result of a single germ cell splitting up into two functionally distinct cells both carried by the same

self-fertilizing organism (full hermaphrodite); subsequently hermaphrodites gave rise to sexes.

4. Crossing may involve many sexes. By analogy with the separation of power in the social domain, there could be at least three sexes: the first is a counterpart to the legislative branch, the second to the executive branch, and the third to the judicial branch. The last sex would oversee that the programs introduced by the other two sexes agree with the fundamental program of development. This prevents the creation of organisms that fail to conform to the basic precepts of development. Before the third sex is found in nature, one could set up and observe the results of a computer simulation of the evolutionary process under the assumption of trisexual reproduction.

5. It is quite feasible that in reproduction by mating the male expresses primarily the end phase of the process of development (i.e., he is the vehicle of environment-induced changes). The female sex is primarily involved with the beginning phases, i.e., profound changes in the structure of the organism. This does not preclude each sex from incorporating functions specialized in by the other sex. Perhaps the above hypothesis ought to be rephrased: Why is it that among developed animals, the testicles are outside the body and the ovaries are hidden deep inside under the skin?

6. Assuming for the moment that females bear deep changes from the beginning and males drive end-induced current changes would seem to imply that in males the germ cells are quite important in implementing change and change in somatic cells is more of an anachronism. Perhaps the mechanism of change in somatic cells is more pronounced in the females. It grapples with on-going environmental fluctuations, freeing the ovum to handle more systemic beginning-induced transformations.

7. The difference between the somatic mechanism of change in males and females points to the possibility of somatic changes being transmitted to germ cells (with the hereditary potential). What lends plausibility to this hypothesis is the following unexplained fact: there exist powerful barriers in the way of foreign cells penetrating the scrotum but none for the ovaries.

8. I would like to note that my comments regarding male and female gametes in no way contradict the theory of sexual types proposed by Lawrence Hurst and William Hamilton. However, the differences in the gametes pinpointed by these authors were used to explain in rather general terms the emergence of different sexual types over the course of evolution. In my opinion, these differences in gamete behavior represent a

subordinate evolutionary trait which makes genetic crossing more effective.

## NOTES TO CHAPTER 7

[1]. It seems to me that the greater activeness of the males as compared with females, a widespread phenomenon in the animal kingdom, is determined by the difference in the production of gametes. Females are basically born with a certain number of eggs (ova). In principle, an ovum can produce an offspring without male participation. Ova are fertilized and can be relatively passive, theoretically always ready to bond. It is another matter that females can pick the best time for mating and under certain conditions reject mating attempts on the part of the male (for example, a pregnant bitch refusing a male dog's overtures). Males, on the other hand, produce sperm on a continuos basis. Only upon reaching a certain level of sperm content is the male ready to actively seek out a mate. In more developed animals the ability of the male to mate is also controlled by erection. Perhaps, this feature is aimed at preventing males lacking a sufficient amount of sperm from attempting to mate since intercourse is so attractive in itself.

[2]. Mating does not imply that the progeny will consist of just two sexes. The reproductive process may give birth to many different types of organisms that do not partake in mating but have special functions in the community. For instance, "The queen honey bee is inseminated by a male just once in her entire lifetime, during the "nuptial flight." The sperm she receives are stored in a little pouch connected with her genital tract and closed off by a muscular valve. As the queen lays eggs, she can either open this valve, permitting the sperm to escape and fertilize the eggs, or keep the valve closed, so the eggs develop without fertilization. ... The fertilized eggs become females (queens and workers); the unfertilized eggs become males (drones)."[204] p. 609.

[3]. I first made my ideas on multi-sexual reproduction public in Moscow at a "duck dinner" organized by a wonderful woman Galina Karpelevich-Poliack. These lines are dedicated to her.

[4]. The exposure of the male organ (partially the female one also) is particularly dangerous when walking erect. Perhaps the Biblical legend concerning the ramifications of Adam and Eve tasting the fruit from the tree of knowledge was fueled by these considerations. "And the eyes of them both were opened, and they knew that they were naked; and they sewed fig-leaves together, and made themselves girdles." (*Pentateuch*, Genesis, 3:7).

Interestingly this legend leads me to speculate that the cardinal difference between man and animal is not really the making of tools (some animals have tools also) but the invention of artificial things to cover his body as called for by changing circumstances (in this case walking erect).

[5]. It seems an experiment of this sort has been performed on real human beings. I want to recount a story told to me by a Russian doctor who came to the US. The doctor was unfamiliar with my hypothesis regarding the mechanism employed by the two sexes to fulfill their function. The story has to do with the aftermath of the

Chernobyl accident. Men who happened to be in the area affected by radiation, but not in the epicenter, and who then left the area and got married to local women had a statistically significant percentage of deformed children. Women who happened to be within the affected area, but not the epicenter, and who then left the area and got married to local men had a significantly lesser percentage of deformed children.

I have merely relayed a story as told to me by one doctor. I have no documented evidence to corroborate these findings. In fact, I could have been blinded by my subconscious desire to hear evidence supporting my hypothesis and this could have distorted my perception. As my now deceased friend, mathematician Boris Moishezon liked to say, if you really want to prove a theorem and its seems like you have succeeded in proving it, the proof is probably flawed.

There is major work being conducted in the former Soviet Union to study the aftermath of the Chernobyl disaster. Foreign scientists have access to the area and scholars interested in the subject could verify the story told above. Even if the results prove negative, the findings of such a study could be quite informative.

[6]. Based on the mutation rate of sperm and eggs in different species, W.-H. Li, *et al.*[205] conclude that the evolution of DNA sequences is driven by the male sex. From my perspective this conclusion is rather rash. There are mutations and there are mutations. Perhaps the authors' approach is valid for minor mutations whose number far exceeds the number of major mutations; the latter are perhaps driven primarily by the female sex.

[7]. It is well known that there are four types of viruses produced by a combination of DNA and RNA on the one hand, and a single or double spiral on the other. Spermatozoon is a kind of virus which has DNA and double helix structure.

[8]. The impulse that inspired me to address this question was provided by Natalie Angier article.[206]

*PART THREE:*


*The Evolutionary Mechanisms of Change: Pathology*


Many years ago I read a popular book from the "Eureka" series published in the former Soviet Union. Unfortunately, I do not remember the name of the author or the title or even what the book was about. But one thing I recall is my fascination with the author's approach to Parkinson's disease. As I recall, the author posed the question in the following way: what is the use of hand tremor under normal conditions? His answer hinged upon the way the hand operates under the conditions of uncertainty: tremor allows for greater flexibility, i.e., the ability to change the position of the hand quickly depending on the immediate circumstances. (Other examples of the same phenomenon - tennis players expecting a serve, goalkeepers in soccer waiting for a penalty shot, etc. - have confirmed the expediency of the hand tremor mechanism when it functions properly). The author then proceeds to discuss the pathology of the tremor mechanism which, he claims, manifests itself in Parkinson's disease.

CHAPTER 8

# FIRST STEPS EN ROUTE TO A NEW CONCEPT OF CANCER

## 1. PATHOLOGY AS A DEVIATION OF AN OTHERWISE NORMAL MECHANISM

I found out afterwards that a number of the more creative biologists and medical scholars have stated this principle explicitly: a disease represents a pathology of some normal mechanism.

Let me quote the following passage from a book by Yury Vasiliev and Izrail Gelfand

> "In biology pathological disorders are frequently studied to probe into the respective normal mechanisms: the study of mutants suffering from metabolism disorders reveals normal stages of the metabolic process; the study of individuals with functional brain disorders pinpoints the location of certain functions in the healthy brain, etc. Similarly, by exploring the neoplastic cells we may come to understand normal interaction between the cells and the environment, normal mechanisms of morphogenesis, proliferation, and differentiation. This kind of research is very complex for it addresses the most intricate mechanisms of cell behavior. We are now just beginning to probe into the nature of cellular response and to develop an appropriate language to describe these interactions."[207] (p.3)

Having recognized this new approach to Parkinson's disease, I asked myself the following question: "Which normal mechanism, of which cancer is an extreme manifestation, is broken-down when the disease strikes?" Unlike Parkinson' disease or other diseases which affect specific functions or parts of the body, cancer in the most developed forms envelops the entire organism.

Of course, cancer may be localized and destroy the organism by striking at individual organs without ever metastasizing (or before it has time to do so??). But to repeat myself, the developed form of cancer envelops the entire organism. This makes cancer a systems disease, a statement which I shall attempt to clarify later on.

Actually, some medical professionals purport that any serious illness is caused by the general state of the organism. This approach shifts the primary focus from the specific ailment to the "holistic" treatment of the body. An original view of holistic treatment has been proposed by Vladimir Dilman.

He attempts to link systemic internal disorders in the organism with the process of aging and specifies ten major non-infectious diseases which accompany aging, including arteriosclerosis, cancer, high blood pressure, and diabetes.

Dilman notes: "Thus, the same mechanisms are responsible for all of the ten major diseases of the middle-aged and elderly. These diseases are so closely interrelated that it is natural to ask the following question: are there really ten individual major diseases, or is there one integral disease with ten major symptoms?"[208] (pp.374-375)

Research on these ten major disorders should not compete with the study of each one individually. In other words, it takes a combination of general and specific factors, both being necessary conditions, for a major disease, or disease in general, to strike.

While Dilman's analysis of diseases is highly interesting, I would like to approach disease classification in terms of my ideas regarding the mechanism of change. One of my quasi-hypotheses regarding the biological mechanisms of change postulates two classes of diseases: one entails disorders in a normally reproducing cell and the other - disorders in the normal mechanism of change. The first group certainly includes such disorders as mechanical damage to the cell. The second class of diseases is more complicated. Based on my classification of the mechanisms of change as they perform over time (section 3, Chapter 3) it is quite plausible that mechanisms aimed at short term changes may cause problems that will manifest themselves in the form of various ailments, but ailments which the organism will be able to overcome. The best example of this case is seasonal virus infections caused by the need on the part of the organism to adapt to the new weather conditions. But even in this case, pathological phenomenon may flare up, e.g. when some dysfunction in the mechanism of change is so severe relative to the organism's immune system as to cause its demise; flu epidemics accompanied by high fatality rates.

Mechanisms aimed at mid-term changes may produce painful side effects since the new things produced may hinder the organism's performance. For the most part these side effects are not fatal and are limited to specific organs, such as cysts or deformation of a single tissue. However, when these changes cause complete deterioration of a given organ or are fatal then we have a pathology of the mechanism of change.

Finally, mechanisms aimed at long term changes may cause severe problems because all kinds of inconsistencies arise as changes in different organs are being mutually adjusted. In the pathological case these changes may be so extreme relative to the immune system's capacity to resist as to lead to the organism's destruction.

We might conjecture that the various forms of cancer, in terms of the duration of the process, are linked with certain dysfunction in the mechanisms of change aimed at temporally different goals.

Whatever the source of ailment, one method for uncovering the specific causes of a disease is to investigate the function(s) performed by the damaged mechanism under normal circumstances. One could then try to eliminate abnormalities in these specific causes since even a temporary relief could be sufficient to ameliorate the disease.

So, I asked myself which specific functional failure could be linked to cancer? One feature peculiar to cancer is the change which takes place in the cells and the propensity of these cells to conquer the organism, including its reproductive system. This led me to hypothesize that cancer might represent a pathology of some mechanism which fulfills the function of change in an organism. By analogy with economics which encompasses the fast growing sphere of R&D with a multitude of respective mechanisms, I thought of an organism as also having various internal mechanisms of change. An organism, just like a society, integrates the sector of change with the production sector, the latter being directed at sustaining and reproducing a given entity. A break-down in the sector of change might cause great harm, even destruction, to society as a whole. Similarly, abnormalities in the organism's mechanism of change, whatever the state of its "production sector", perhaps, could be quite detrimental.

In exploring the various directions to develop my hypothesis, I decided to start by examining the normal operation of a rather intricate inner mechanism of change based on somatic cells. I soon realized that there was no relevant biological theory which would be acceptable to the scientific community and which would satisfy the rigid standards of empirical science. However, there is very little empirical material on the somatic mechanism of change as it operates in the norm. I could summon a few works to corroborate certain isolated parts of my theory, but they would not add up to a coherent scientific hypothesis. In those rare instances when reputable scholars did voice their views on the subject, their ideas were rejected outright because they are associated with the heretical idea that somatic changes are heritable or beneficial in terms of adaptation. All these theories smell of obsolete Lamarckism, not to mention the fact that the aversion to any hint of somatic heredity was fueled by charlatans exploiting this idea.

Due to this deficit of scholarly material I saw no way to develop directly a coherent theory of a normal mechanism of change which I imagine to exist. Meanwhile, research in molecular biology has produced a vast amount of data describing a pathological mechanism of change.[1] So, sticking to my main hypothesis, I reversed the course of my investigation and corroborate the existence of the various components of the mechanism of change by showing their pathological expression. I realize this approach is controversial but it is not without merit. The general normal mechanism thus reenacted could provide clues in understanding the intricacies of its pathological manifestations. I shall illustrate this methodology with numerous examples dealing with cancer research.

So, cancer, as I define it, is basically an extreme form of a pathological attempt to restructure an organism using the mechanism of somatic change.

## 2. CONTEMPORARY VIEWS ON CANCER

### 2.1. Universality of Cancer

Assuming cancerous diseases derive from a pathology of the somatic mechanism of change then this class of disorders must encompass all developing life forms from plants to humans.[209]

Some scholars[210] deem that cancer is not universally intrinsic to all life, allegedly sparing plants. (Some reports[211] claim only limited similarity between plant and animal cancer). However, there is much evidence that points to the contrary. Noteworthy in this respect is the general overview of the subject containing the following passage (I left all references to demonstrate the interest generated by this topic).

"As just one example we want to concentrate on a kind of cancer which is observed in different kinds of plants (dicotyledonous), e.g. sunflower or tobacco [M. De Cleene and J. DeLey, *Bot. Rev.* 42, 389 (1976).], namely the crown galls. The interesting fact is that these oncogenes are of bacterial origin, i.e., they are a unique example of the expression of prokaryotic DNA sequences by eukaryotes. The bacterium which is able to transfer carcinogenic plasmid DNA into the genome of plants is called *Agrobacterium tumefaciens* [W. B. Gurley *et al*, in *Genome Organization and Expression in Plants*, C. J. Leaver (ed.), Plenum Press (New York, 1979), pp. 481.]. The plasmids which induce the cancer are ring shaped DNA molecules [Zatnen, I. *et al*, J. Mol. Biol. vol. 86, 1974, p.109.] and are different in different tumors, however, they contain as one part a special sequence of DNA which is the same in all cases and is called T-DNA [Chilton, M. *et al*, in *Genome Organization and Expression in Plants*, C. J. Leaver (ed.), Plenum Press (New York, 1979), pp. 471.]. Most probably this is the cancer causing sequence, i.e., the oncogene.

The origin of cancer from the ring shaped so-called Ti-plasmids was proven by the observation that removal of this plasmid from oncogenic strains of *Agrobacterium tumefaciens* results in a loss of oncogenicity, while the introduction of this plasmid into the genome of non-oncogenic strains of the bacterium produces oncogenic forms [Van Larebeke, N. *et al*, *Nature* vol. 252, 1974, p.169 ; Watson, B. *et al,J. Bacteriol.* vol.123, 1975, p.255.]. Obviously we have here an example of an early discovered oncogene of non-viral but bacterial origin in plants. A review on this topic can be found in [Schell, J. and Van Montagu, M., in *Genome Organization and Expression in Plants*, C. J. Leaver (ed.), Plenum Press (New York, 1979),pp. 453 and other papers in this volume.][212]

Cancer has been discovered in animals living long before the advent of the human species. For instance, in the beginning of the 1920s bone tumor was discovered in a dinosaur skeleton found in Wyoming. Bone tumor was determined in a 1,500,000 year old human jaw and signs of malignant bone tumor were discovered in the septum hole of a human skeleton dating to the Neolithic period. Bone cancer was found in an Egyptian mummy.

Papyrus and other documents unearthed in the Middle East allude to tumors of the various organs. Cancer is mentioned by Roman, Greek, and Jewish sources. References to cancer have grown steadily reaching astronomical proportions with recent progress in medicine and biology.

It should be noted that animal and plant cancer are not completely unrelated. The work cited below presents comparative analysis of abnormal processes of growth in plants and animals examining neoplasmatic (epigenetic) and neoplastic transformation as well as hereditary differences in growth parameters of these two kingdoms. First proposed 80 years ago, the affinity between plant and animal cancer was subsequently rejected, but as a result of some recent findings is making a comeback.[2]

Cancer also encompasses a variety of inner structures at the level of individual organs as well as tissues and cells. More will be said concerning the problem of organs that are known to be susceptible to cancer. For now I will limit myself to a few preliminary remarks regarding cells and tissues.

Cancer is known to affect many different tissues and cells. This point is important since one facet of my hypothesis of cancer concerns the process of change as it unfolds in any tissue or cell unless its growth is halted. There is no need to dwell on the susceptibility of the various cells of the epithelium tissue to tumor. Just recall that muscle tissue (smooth and striated), although less prone to cancer, is susceptible to both primary and secondary tumors, some of which are malignant and can metastasize.[213-217]

## 2.2. Empirically-Driven Quest for the General Theory of Cancer

Today we have no general theory of cancer which would consolidate the latest discoveries in molecular biology. One indirect confirmation of this theoretical vacuum is the multitude of mainly descriptive definitions of cancer focusing upon some specific aspect of the disease or its causes. (These definitions are discussed in detail below.)

The rather simplistic empirical approach to cancer really goes hand in hand with the recent progress of the last three decades in molecular biology that has raised cancer research to a new level. It seems that most scholars believe that the only way to deal with this terrible disease is to uncover at the molecular level the underlying cellular mechanism.

Many biologists claim to understand the sequence underlying the malignant process. Moreover, in order to stop the disease they deem it sufficient to eliminate one necessary condition underlying tumor growth. In other words, they think the process can be stopped at any stage before it destroys the organism. Outstanding achievements made in this branch of cancer research serve as a foundation for in-depth analysis and prevent superficial and unsubstantiated speculation.

Advanced research in this area has made it possible to fight cancer by impeding its progress at the various stages of the disease. This has lead to the idea entertained by many biologists of stopping cancer by thwarting telomerase activity which excites telomeres which, in turn, results in unchecked growth of cells. Other methods include revitalizing genes responsible for slowing down cell growth, stimulating repair mechanisms in the cell, and preventing the formation of tumor infrastructure, especially the blood supply system.

I would say, following the terminology used by John Wyke, that the recent advances in cancer research have evolved primarily from the "bottom up."[218] While there has been some progress in the "top down" approach, the main thrust of the research has been put at the intermediate levels of the hierarchy, for example, the structure of the tumor. So, in spite of this vacuum in general theoretical framework, outstanding experimental discoveries of recent decades have revealed the intricate workings of cancerous cells at the molecular level.

Considering the organism's physiological complexity, it might seem unreasonable to distract scientists from conducting new and valuable research into the operation and interaction of cells in favor of general theorizing about cancer. This attitude to practical vs. pure theoretical research implicitly presupposes that once the sea of empirical information becomes incomprehensible making further experimental data intractable, scholars would have no problem concocting the necessary theory to organize the data.

However, the history of science has proved otherwise. For example, in physics a leap into the subatomic realm was accompanied not only by pioneering developments in physics itself (such as quantum mechanics), but also have been preceded by new currents in philosophy, particularly the advent of positivism. Physics was one area where theory and practice worked in unison to complement each other.

I believe that the present day shortage of elaborate hypotheses regarding the nature of cancer as a systems phenomenon represents a major obstacle to further progress in this field. I hope the theory of cancer expounded here will help remedy the situation.

## 2.3. A Systems Approach to the Definitions of Cancer

Just like any definition, a definition of cancer is neither exclusive nor exhaustive.[219] Rather than try to come up with such an exclusive and exhaustive definition, a pluralistic mechanism of definition construction and selection is introduced in Chapter 3 of this book.[220] First and foremost in this mechanism is the construction of many definitions; we then proceed to select the definition which best serves our immediate pragmatic goals. The third step is to determine whether or not the selected definition is really expedient, and finally the selected definition is replaced by another definition in case the original one proves it is inefficient or inadequate in terms of the changing conditions.[3]

Definition of cancer from the systems perspective can be given in terms of function, structure, process, or genesis. All four dimensions are mutually complimentary.

From the functional point of view, the definition of cancer must specify its function. Some biologists hold that from the functional standpoint cancer plays the role of weeding out the weak; the mechanism is perhaps triggered automatically when a person reaches a certain stage. This would, for example, explain why cancer tends to strike older people. However, this approach raises a very reasonable question: why introduce such a complex internal mechanism of purging the old, a mechanism which not only destroys tissue where cancer originates (in itself sufficient to weed out the old), but does not direct the metastasis into organs whose quick failure would be lethal. We know that physical decline and death can result from a much less intricate internal process of degradation, not to speak of external factors such as viruses, microbes, and predators which are quick to strike the weaker and older organisms.

Even if we accept for the moment this functional definition of cancer, the question of the precise weeding out mechanism remains open. An interesting perspective on cancer from the functional point of view portrays it as a peculiar catalyst in the process of mutation.

> "The highly structured mechanisms of cancers, their tendency to occur as a response to environmental stress, and the existence of oncogenes, suggest that neoplasticity may

represent more than a biological dysfunction. It is proposed that cancer exists as a phylogenetic mechanism serving to promote "hyperevolution", albeit at the expense of the ontogeny, that is similar to a process recently discovered in bacterial mutations. Cell-surface-associated nucleic acid in tumorigenic cells and sperm cell vectorization of foreign DNA indicate the existence of essential mechanisms necessary to the occurrence of cancer mediated hyperevolution. An analysis of the proposed mechanism indicates that for mutagenesis of chemical cytology, stress induced neoplasticity confers an evolutionary advantage of more than two orders of magnitude."[221]

I share the view espoused in the above passage that cancer is linked to the mechanism of generating new mutations. However, taken in isolation this factor obscures the more specific role of cancer as a pathology of the somatic mechanism of change.

Many more common definitions of cancer are based on the structural or process-oriented outlook.

Structural definitions of cancer focus on the various components of cancer and their interaction. The common thread running through these definitions is the presence of radical tumor-forming cells and the immune system's ineptitude in restraining them. The following definition is an example of the structural approach: "Cancer is a malignant neoplasm." [222]

Process-oriented definitions emphasize the progress of cancer and its uncontrollability, i.e., the idea that cancer represents unorganized multiplication of cells. In keeping with our methodological device of uncovering the nature of the disease by reconstructing the respective normal but malfunctioning mechanism, cancer represents a break-down of the mechanism normally responsible for managing cell growth. However, these definitions fail to account for such typical phenomenon as metastasis. Biologists insist (see the hypothetical stages of somatic change described in the previous chapter) that the normal mechanism of cell growth does not exhibit migration of cells from one organ to another which would be reminiscent of the metastasizing process.

Here is a typical process-oriented definition of cancer: "**Cancer** a general term for more than 100 diseases characterized by the uncontrolled, abnormal growth of cells in different parts of the body that can spread to other parts of the body."[223] (p. 1)

Essentially similar definition is given in Webster New International Dictionary "Cancer. A malignant growth of tissue usually ulcerating, tending to spread by local invasion and also through the lymph and blood stream, associated with general ill health and progressive emaciation."

These definitions are not very precise. Jumping ahead I want to note that cancer cells are not necessarily uncontrollable nor is their growth unregulated. Reiterating my analogy with dissident radicals in society, radicals include both anarchists as well as communists, fascists, etc., revolutionaries who try to impose their rigid program for governing a society - a relatively superficial program given the complexity of the issues.[4]

Another example of a process-oriented approach to cancer is the following descriptive definition which emphasizes the progress of the disease: "Cancer. A malignant growth of tissue usually ulcerating, tending to spread by local invasion and also through the lymph and blood stream, associated with general ill health and progressive emaciation."[224] (p. 55)

Consider the following definition of cancer from the process-oriented point of view: "Cancer...is a disease in which individual mutant cells begin by prospering [selfishly] at the expense of their neighbors but in the end destroy the whole cellular society and die."[225] (p. 188)

The above description encapsulates the idea that cancer cells behave as *egoists* who pursue their own interest ignoring the interests of other cells and eventually perishing themselves. I believe the definition would be more constructive if it made an attempt to distinguish between cancer cells and innovator cells; at first glance the latter exhibit similar behavior in pursuing their own interest and seizing additional resources. However, innovator cells are individualists, meaning they pursue their own interest while taking the needs of other cells into account.

In terms of genesis, definitions of cancer pinpoint its sources, such as carcinogenic substances, radiation, viruses, etc. as well as its general evolutionary origins.

All in all, in terms of etiology the predominant view today is that cancer is a result of external factors such as carcinogenic substances, radiation, etc.; as a result the thrust of the struggle against the disease is to eliminate these factors. However, an opposing theory that has been gaining momentum asserts that the primary cause of cancer is the organism's predisposition toward the aforementioned factors.[5] In other words, in the absence of other factors, carcinogenic substances are generally not sufficient to produce a cancer cell. Perhaps, these other factors are expressed as modified genes in the DNA program triggered by dysfunction in the higher-level program.

In introducing the agenda of the annual meeting of the American Society of Clinical Oncology that took place in Dallas in 1984 Sandra Blakeslee writes:

> "The new research shows that cancer is not primarily caused, as many Americans tend to think, by the poisons spewed into the air, water, and land by uncaring industrialists. Rather, each person is born with various genetic susceptibilities, essentially weak spots in their genetic makeup, that play a leading role in the cellular mayhem called cancer.
> For example, researchers have found that some people have genes that enable their bodies to detoxify chemicals rapidly, including the carcinogens found in cigarette smoke and natural carcinogens found in foods. Others are born with slow acting varieties of the same genes; their bodies are less efficient at getting rid of carcinogens. If exposed to large enough quantities of the chemicals, these slow detoxifiers are more likely to get cancer."[226]

In the light of the considerations expressed above, the role of heredity and the related germ cell mechanism in causing cancer. The heredity factor comes in if only because the mechanism of change in germ cells has turned pathological in parents. As a result the children inherit a damaged program of development (or the program that changes this program). Consequently, as the zygote develops the fetus assimilates a destructive program that is implemented via the fetus' somatic cells. This might be the reason why doctors administering preventive treatment try to find out about the history of the disease among the patient's blood relatives, the link between the somatic and germatic mechanisms of change in parents who never had cancer. In this case one can assume that the basically healthy predecessors have slowly, over the course of many generations, been accumulating some kind of predisposition toward pathological operation of their germ cell structure without these changes being physically expressed. However, at some point the pathological changes ripen sufficiently to produce a generation of children afflicted by cancer.

The following quotation is interesting from the standpoint of cancer's genesis :

"If evolutionary theory is modified to include the assertion that cancer established, about 700-800 million years ago, the imperative that only those Bilaterian genotypes capable of extreme precision in the construction of multicellular organisms could possibly survive to participate in the struggle for existence and ruthlessly enforced that imperative ever since, then evolutionary theory is strengthened." [227]

This is an attempt to link cancer with the development of complex organisms. My hypothesis, however, that cancer represents a disturbance in the overall evolutionary process is equally applicable to complex as well as simple living creatures. In other words, if there exists, at the level of the organism, an orderly evolutionary mechanism of change susceptible to malfunctioning, there is a likelihood of cancer. Admittedly, the more complex the organism, the greater the likelihood of breakdowns in its evolutionary mechanism, and hence of cancer.

To sum up, disorders in the biological mechanisms of change are fraught with grave consequences. In a sense the most powerful constructive forces underlying development join with highly destructive forces also incorporated into the development process. The major threat stems from cancer dispersion when disorders are not localized to their place of origin. Naturally, the more important the given mechanism of change to the species' development the more grave the implications of its break-down. Disorders in these mechanisms in complex organisms having an intricate mechanism of change (especially if active) can lead to major diseases such as cancer. [228]

## 2. THE PROPOSED CONCEPT OF CANCER

### 2.1. Change and Cancer

The material in the previous chapters was designed to lead to the following general question as a starting point of our discussion of cancer: "What normal biological mechanism leads to this devastating disease?" It was suggested that the answer be sought in the break-down of the somatic mechanism of change instilled by evolution. It is upon this substrate of somatic cells that the tragedy called cancer is being staged. Somatic cells are sufficient for *cancer* to take root provided at least one or , in the spirit of the multi-factor concept of the etiology of cancer, a combination of the following conditions is satisfied.

First of all, damage in somatic cells is induced directly by external factors outside the organism (carcinogenic substances) as well factors inside the organism (viruses).

Secondly, the somatic cells possess internal mechanisms of change that are susceptible to pathological disorders.

Pursuing the latter venue leads me to hypothesize that the genetic structure of somatic cells, apart from carrying genes responsible for their own reproduction, includes genes responsible for change and that these genes can, under certain conditions, become oncogenes. One piece of evidence supporting this claim is the discovery of highly repetitive sequences of genes, called minisatellites, in the structure of the "selfish genes"; these mutations interfere with growth regulation and induce some forms of cancer[229]. In fact, assuming the internal mechanisms of change reside in the "selfish genes" (see chapter 5), the above hypothesis is further strengthened by the fact that the structure of "jumping" genes and certain cancer-causing viruses is identical.[230]

I will now try to conceptualize the phenomenon of cancer with the help of the systems approach.

## 2.2. Systems Approach to the Theory of Cancer

Currently cancer is viewed as an event that takes place in an organism (total system) whose function in the norm is *survival*. Faithful to the guiding principles of the systems approach with its emphasis on the system (organism) as a whole, in which the target object is immersed, I have proposed a new paradigm for probing into the phenomenon of cancer. It is predicated upon the idea that the function of an organism is not limited to reproducing creatures in its own image. Organisms aim at *development* (change). I further surmised that complex systems, to which living creatures belong, ought to include more than one mechanism of change. Beside the widely eulogized mechanism of random mutations, I believe there exist diverse mechanisms of change that are ordered, to a greater or lesser extent.

To reiterate, I have focused on the somatic mechanism of change in multicellular organisms possessing rather advanced intercellular communication network. Molecular research into the structure and development of normal and cancer cells has shed much light on normal as well as malignant processes in the organism. As a result, the notion that genes present in the normal cells may, under certain conditions, turn into oncogenes has gained general recognition.[6]

What's more, this proposition holds with respect to not just the normal cells, that is cells that merely reproduce a given organism, but to innovator-cells as well.

Here I would like to remind the reader of James Shapiro's observation: most biologists regard as *normal* only those cells which reproduce a given organism. They regard changes in the cell as a result of some kind of damage (break-down, mistake), i.e., changes in the genome are regarded as a deviation from the norm and always bear a negative connotation. This approach to genetic change agrees with the prevailing views on evolution as being driven by random mutations resulting primarily from external fluctuations. Accordingly, all afflictions of the genome are manifestations of these random mutations.[7]

Thus, any cell whose structure deviates from the norm is often labeled damaged, i.e., abnormal. Shapiro notes, however, that a change in the cell could be a normal phenomenon which serves to promote the cell's adaptation to its new environment.

> "Molecular genetic results have tremendously expanded our understanding of what living cells can do with their genome. The examples described above illustrate some of the many ways that different biochemical systems serve to restructure DNA molecules in organisms as diverse as bacteria and mammals. These DNA reorganization systems are subject to cellular regulation, and some of them serve specific adaptive functions in organismal life cycles. It is easier to understand how change can be regulated and used to meet adaptive needs if we think of it as a biochemical process rather than as a blind consequence of physico-chemical damage. Such damage does occur, of course, but it is anticipated, and the contribution of purely chemical events to genetic change is kept at a very low level by elaborate repair systems."[231]

It can further be surmised, and there is plenty of evidence to support the claim, that organisms possess special features to remedy various breakdowns of some normal mechanism that is subject to failure. In other words, once the flaws in some functionally useful mechanism are uncovered, meaning we have determined the nature of the disease, the organism's own powers should be summoned to shield it from the disease.[232] Of course, if the organism's inner resources prove inadequate, thus creating a pathological situation the organism perishes unless it receives timely outside help.

All this leads us to an important distinction between *normal* and *pathological* break-downs. Both cases may lead to a disease, but normal flaws, unlike the pathological case, can be overcome by the organism itself.[233] Pathological disorders vary in time and in space in the extent of havoc they can wreak, progressing slowly or rapidly, attacking a single organ (part) or many organs, changing or destroying them. For example, the simplest form of pathology entails degeneration when the cell (tissue, organ) begins to disintegrate, reverting to its early state that is unstable in the new environment. Of course, degeneration may assume extreme forms when the organ is ravaged completely. Another form of pathology with less severe consequences is transformation (reconstruction) of the cell manifested in benevolent tumors, deformities, etc. Still, there is a big gap between these changes in the cell and cancer.[8]

The more extreme expressions of pathology engulf many organs and destroy them very quickly relative to the life span of the organism.

Under certain conditions pathology can be treated by the rebirth of the cell, i.e., the cell's reversion to its original state. However, rebirth is not a universal remedy because the original state is predisposed toward all kinds of abnormal phenomenon. The other method is cell *rejuvenation* which entails elimination/introduction of certain elements.[9]

Particularly interesting in this connection is the relation between *apoptosis* and cancer. If we regard the former as a normal phenomenon then cancer could be deemed its pathology in the following sense: cancer leads to the rejuvenation of a cell that was about to die a normal death. Research conducted in the 1980s has proved that the oncogene *bcl-2* introduced in cells undergoing apoptosis results in cell survival.[234]

Correlation between apoptosis and cancer parallels in some respects that of telomeres and cancer. Typically cancer cells revitalize telomeres whose absence would otherwise be fatal (see section 2, Chapter 6). In my scheme of cancer, cancerous revival of a doomed cell is the work of the mechanism of change which must, before all else, ensure that cell's survival. Dysfunctions of this mechanism, which in itself are the cause of cancer, result in this paradoxical alliance between the cell's rejuvenation and such lethal agent as cancer.

Let us examine the behavior of normal and pathological cells in the light of our discussion of social systems in Chapter 2.

## 2.3. "Deviant" Cells and Cancer Cells

All cancer cells are deviants. They are characterized by 1) progressive multiplication, 2) mobility, 3) diminished adhesiveness (loss of cohesion), 4) phagocyte activity.

Generally speaking these features are shared by all innovator cells. Some characteristics of cancer cells may be missing in innovator cells, but these characteristics may not be universal for all cancer cells either. To continue with the above list of features we could add 5) the output of these cells may contain poisons - tumor metabolites.[235]

The critical question is how to distinguish between innovator cells and cancer cells. It is important to recognize a cancer cell at an early stage of its development. Diagnosis based on the physical expression of the deviant cells may be too late in terms of stopping the cancer cell from destroying an organism. Is an early diagnosis possible?

The answer is not that simple. The astounding scientific and technological strides of seventeenth-eighteenth centuries made scientists and engineers believe that virtually everything is possible from perpetual motion machine to utopias (paradise on earth). The nineteenth century was more sober. It began with the discovery of the laws of thermodynamics proving the impossibility of perpetuum mobile. Soon after the great mathematician Evariste Galois proved that equations of the degree higher than four have no general analytical solutions. Other examples of impossibility abound. In the twentieth century this category of proofs extends beyond mathematics (Kurt Gödel's theorem) and physics into the realm of social sciences (Arrow's theorem proving that under certain conditions no solution to democratic voting procedure exists).

In solving complex problems modern science first attempts to prove that the solution exists and then proceeds to find it.

With respect to an early diagnosis of cancer the question is can we formulate sufficient conditions to distinguish a cancer cell from other deviant cells. Of course, at the present time the possibility can neither be proved nor refuted. I am rather skeptical; it is quite plausible that innovator and cancer cells are indistinguishable at an early stage of development. It may be discovered sometime into the process that an innovator cell has turned into a kind of "killer on the loose" which strikes haphazardly or "killer-robber" which kills selectively in pursuit of his own gain. Moreover, innovator cells may include various brands of "radicals" including "terrorists." Rather than invigorate a complex organism these cells try to subordinate it to their dictate based upon a superficial scheme with which the immune system is unable to cope.

It is especially difficult to stand up to the radicals in times of crises (in a sense similar to an organism under stress) when they promise to alleviate the current hardships. The tricky part is that the measures they propose may prove successful but only ... locally (in the short run). This was the case with fascists in Germany. Having gained power they reduced unemployment, raised the standard of living, returned Ruhr territories lost after WWI back to Germany, and so on. At the same time the fascist program contained seeds of the terrible things to come. Few were really concerned because the majority focused on the current hardships and short run relief. It took American political culture not to succumb to communist or fascist (corporate socialism) ideologies at a time of crisis while borrowing certain ideas from these groups. Interestingly, Keynes was compared to Marx as far as augmenting government power, and Roosevelt was regarded by some as "pink."

By analogy with social systems it is not necessary for cancer cells to be unmanageable or exhibit profuse multiplication. Their behavior could be organized by being governed by their own destructive program. Coupled with the inability of the immune system to channel the resulting changes along a desirable course, the consequences of cancer are devastating.

My speculations, assuming they make any sense, lead to some practical, albeit remote, conclusions regarding our attitude toward cancer.

Today the medical profession does its utmost to destroy cells deemed malignant or restore them to a "normal" state in the orthodox sense of the word, i.e., a conformist cell that reproduces a given organism. The last method is summoned to fight cancer by deactivating telomerase activity in cancer cells, i.e., restoring these cells to their normal cycle with a prescribed number of divisions and no side-affect on the healthy cells. "The paradigm is we have a new way of dealing with the tumor cell, not by killing it or poisoning it, but by reasoning with it."[236]

Since innovator cells may behave similarly to cancer cells, especially at an early stage, it is not advisable under early diagnosis to categorize all *innovator cells* as cancerous. This confusion in cell classification may hinder evolution by depriving innovator cells of opportunities under the guise of destroying cancer cells. I believe deviant cells should be approached in the following manner. Even if the subsequent course of development of radical cells is unknown and there is no guarantee that the immune system will be able to cope with radical cells if they turn out to be terrorists, it may still be advisable "to isolate these cells from society" rather than destroy them. By analogy with radicals in a democratic society, once this kind of group asserts its existence it can be "shadowed" or spied upon. Radicals may be kept away from classified work and their organizations infiltrated by agents. As soon as they turn to an armed conspiracy they can be isolated and tried in court. The most severe punishment is administered only if they initiate some kind of unlawful action.

Similarly, radical cells can be subjected to all sorts of isolation. Subsequently, if the organism proves capable of assimilating these cells they can be allowed to function on par with other cells. A number of oncologists have arrived at the same conclusion although for a different reason. They realize that the destruction of cancer cells entails hazardous procedures including the possible disruption of healthy cells typical of chemotherapy.[237]

I realize full well that at the present time when cancer is still extremely lethal my comments sound rather frivolous (to put it mildly). It seems absurd to think of development when survival is on the line. As the saying goes, "Thank God just to be alive." But thinking of the future (at least occasionally) is bound to pay off. Indeed, destruction of all innovator cells could prove detrimental to the sources of change that give viability to the cell with all the ensuing consequences (such as the organism's life span). This statement is based on the role of telomeres discussed in the previous chapter.

The various "social" functions played by the cancer cells point to a connection between the function and the form of cancer.

Is not sarcoma a form of cancer in which the malignant cells resemble killer cells that simply take away nutrients or space from other cells by poisoning them with the waste they secrete? Less extreme forms of cancer that exhibit slower growth can affect change in the genetic structure of other cells; this resembles the way many radical groups operate. The creeping forms of cancer are akin to "revisionist" type cells (see typology of cell based on the analogy with deviants in society in Chapter 2).

Let us categorize all deviants that inflict harm as "traitors" since at first they all promise prosperity; cancer cells belong to this category.

"Cancer is a kind of treachery from the interior of the body. Cells which have been "good citizens" so far obeying the laws exactly changing their nature, beginning to perform division and hyper plasia indifferent to the demand of the living body, intruding into adjacent normal tissue, besides spreading over other parts of the body, taking root there, and make the second base equally unlawful (metastasis). Such an unlawful castle can give rise to almost all organs in the body."[238]

Our discussion will focus on the "traitor" cells, meaning the general characteristics of cancer, and ignore the peculiarities of the various forms it assumes.

## 2.4. The Possible Usefulness of My Hypothesis on the Nature of Cancer

Giving in to vanity and fantasy, but only to illustrate the value of theory in cancer research, I would say that if a theory like mine was around in the 1960-70s it would have served as a catalyst in research on genetic changes that lead to cancer. It seems that for many years cancer research at the molecular level was focused on finding special cancer genes, or even one all-pervasive cancer gene. It was not until 1976 that M.Bishop and H. Varmus discovered that the gene which was thought to cause cancer in chicks was essentially identical to a normal gene. Thereafter, research shifted from one all-pervasive cancer gene to the underlying causes of disorders in the genetic mechanism at various stages of the disease.

Of course, who knows what could have been? I think a better way to bolster my case is to show the pertinence of my conceptual approach to future research. If nothing else, my optimism is based on the new vision afforded by my approach. It reveals the link between various stages of cancer, the nature of each stage, and the sources that might trigger the malignant process at each stage.

Possible evidence in support of my hypothesis concerning the nature of cancer may be found in the study of certain types of pathologies. Presently, biologists are at work looking for the links between arteriosclerosis and cancer: both conditions feature blockage of gene *p53* which is responsible for slowing down the rate of cell reproduction.[239] In case of arteriosclerosis, however, the rapidly multiplying cells impact the organism only *locally* by depositing themselves on arterial walls and interfering with blood flow. In case of cancer, the effect of rapidly multiplying cells is *globally* in that they gradually spread throughout the organism.

It is quite possible that local organ-specific diseases are pathological manifestations of the workings of stable reproductive mechanisms, whereas systemic diseases, such as cancer, are pathological manifestations of a cell changing its somatic mechanism.

It bears repeating that cells possess a mechanism to repair damage both as to the coding genes and as to the mechanism of change. In cancer cells genome damage is not repaired. One's conceptual framework guides one's search for causes that explain this type of behavior by the repair mechanism. Assuming some changes are of an innovative nature, the repair mechanism need not intervene. At some stages of cell development, cancer cells may behave just like or at least similarly to innovation-bearing cells so the repair mechanism remains idle. On the other hand, if any change in the cell is deemed abnormal - a deviation from some states both as of normal reproduction and as of change, then this passivity on the part of the repair mechanism will always be regarded as a disorder.

It seems to me that the silence on the part of biologists regarding the function of the internal somatic cell-based mechanisms of change and the impact of these changes on heredity is rooted in the reigning doctrine which postulates a rigid chain linking change and heredity: the idea of an internal mechanism of directed change is rejected with positive change being implemented solely via selection among organisms that have undergone random damage to the genetic structure of the germ cells.

I should note that certain ideas presented here concerning the link between cancer and the process of evolution have been voiced independently, although in different and rather general terms, by a number of biologists.

For instance, I found a published report linking cancer and evolutionary change in the following context. The possible connection between cancer and certain genetic diseases such as Huntington's disease, myotonic dystrophy, fragile X syndrome, and spinal and bulbar muscle atrophy has generated some interesting questions. The impetus for research in this area was the discovery revealing similarities in the structure of the genes responsible for these diseases. Genetic affinity is manifested in the abnormal repetition or duplication of the same segment of DNA in the genome. Bert Vogelstein noted that this phenomenon of abnormal duplication is characterized by an instability of the genome as opposed to its stability as a norm. He speculates that "The expanding segments might speed genetic changes that allow an organism to evolve and in the case of cancer this mechanism goes awry." Vogelstein continues, "One way to think about tumors is as an evolutionary process in which evolution occurs fanatically."[240]

However, Vogelstein does not elaborate upon this point and fails to provide any explanation of how such a mechanism of evolutionary change might actually work.

One noteworthy exception to my criticism is the concept of cancer elaborated by Lev Meckler. (A word of warning. I would not venture to judge the validity of Meckler's concept so my criticism will be rather general. According to expert opinion, which is generally skeptical, the published account of Meckler's work represents a peculiar blend of plausible ideas corroborated by credible references, loose interpretation of certain facts, and axiomatic statements of dubious origin.)

According to Meckler, cancer can be viewed as a pathological manifestation of the mechanism of adaptation to changes in the environment. It entails changes in the structure of the organism resembling the process of embryonic development.[241]

For me it is not necessary that the changes induced by the somatic mechanism be transmitted to germ cells; Meckler does cling to this point. The mere existence of this mechanism is sufficient. Perhaps the somatic mechanism is archaic and changes that it produces are not allowed to pass on. In fact, in higher organisms if changes in the somatic cells do reach a germ cell, it is more likely to occur in females.

My other point of contention with Meckler is the source of change. In my opinion, the impulse does not have to come from the environment. Meckler's emphasis is on the organism's immediate response to external demands, i.e., the passive nature of the mechanism of change. The importance of this response mode cannot be denied. At the same time, we should not underestimate the role of the mechanism of change that takes signals from inside itself and *actively* bears down on the environment. This mode is directed at conquering and modifying the world and in those cases where change can be initiated at the beginning it incorporates the tunnel process.

Meckler is very categorical about the affinity between cancer and embryonic cells. I believe he stretches the comparison between the two. The cancer cell transmits change to different organs while an embryonic cell carries the program of development of a given cell to a specific structure of an organism.

Furthermore, if the cancer cell was embryonic its development should not be limited to tumors, assuming the latter is a forerunner of organ formation. It should also start to form organs even if deformed ones.

This leads me to disagree with Meckler's claim that the migration of somatic cells is intrinsic only to pathological cells such as hybrid somatic cells and that information from normal organs is transmitted to other organs solely through the viruses. My concept, to put it boldly, does not rule out the migration of normal somatic cells, which include innovator cells. Therefore, in developed organisms cell migration is not the prerogative of cancer cells. As noted above since the process of somatic change requires cell coordination we should not rule out the possibility of a messenger somatic cell returning to its host organ; these considerations apply equally well to cancer cells. It seems that such reentry is not characteristic of embryonic cells.

Hopefully, the reader is convinced that the field of etiology of cancer leaves plenty of room for new ideas.

## NOTES TO CHAPTER 8

[1]. Sometimes the situation is reversed. In discussing the suppressor genes one author notes: "While there is much to be learned about what the proteins encoded by the suppressor gene do, researchers are making progress toward understanding how these proteins normally function - and also how they may malfunction in cancer."[242]

[2]. "Tumorigenesis in eukaryotic organisms is based on the deregulation of normal cell growth and development. This deregulation may be elicited by external as well as endogenous factors. We distinguish between benign and malignant growths depending on the inducing tumorigenic agents and on the geneticmake-up of the affected organism. This review discusses similarities of neoplasmatic (epigenetic) and neoplastic transformations in plants and animals as well as inherent differences in the growth parameters between the two kingdoms. Examples given for neoplasmatic tissues are the hyperplasias and insect galls (zoocecidia) in plants and hypoplasia, aplasia and agenesis in animals (and man). Neoplastic transformation in plants is the result of either the incorporation of foreign nuclear material into the plant genome or an imbalance of inherited chromosomes (in hybrids). Examples for neoplasias are the crown gall disease and Kostoff's genetic tumors in plants, and the carcinomas and leukemias in animals. The more than 80 year old, but neglected, concept of a correlation between tumorigenesis in animals and plants has been revived through advances in molecular and cell biology and molecular genetics which will stimulate a new form of biological reasoning and thought, fueled by new insights into cellular regulatory processes."[243]

[3]. An extreme example of this approach is the definitive two-volume set on cancer.[244] It simply ignores the issue of definition of cancer. Some mathematicians who want to avoid scholastic arguments define mathematics as "anything that mathematicians deal with". Perhaps, the authors of the above publication had similar concerns.

[4]. Thus, communists-radicals (revolutionaries) were able to seize power in Russia and fascists to do the same in Germany. These groups then managed to impose their rule upon the entire country with various segments of society controlled by these groups evolving in a rather organized fashion. It is quite another matter that their primitive "genetic program" penetrated so deeply into the fabric of the social organism, it distorted the structure of the country and its mechanism of government, and it deformed people to such an extent as to lead the country within a relatively short historic time period to a virtual exhaustion with all the ensuing consequences.

[5]. The study of the regional aspects of cancer distribution has revealed a correlation between nutrition in various countries and different forms of cancer. Perhaps this is not a cause-effect relationship but merely a correlation: the genetic make-up of different peoples may create predisposition toward the various forms of cancer.

[6]. I do not know whether Meckler was the original exponent of the idea that genes present in the normal cells may, under certain conditions, turn into oncogenes, but certainly one of first.[245] In his later works, Meckler summed up his results: "It has finally been determined - contrary to the prevailing theories of oncogenesis - that there is no single transforming gene, or so called "sark" gene. Therefore, there is no single transforming protein capable of transforming any cell. Based on experimental results the theory of oncogenesis espoused here states that genes responsible for cell transformation are the organism's normal genes which differ in terms of their organ or tissue characteristics. These genes are localized in specific sections of the genome that are no different from the sections housing its endogenous viruses, i.e., structures which, according to the present theory of oncogenesis, carry these genes from one type of cell to another."[246]

Whatever the priorities in this field are, in 1989 one of the few Nobel prizes awarded specifically for cancer research went to M. Bishop and H. Varmus. They discovered that normal cells carry genes which, if they malfunction, cause cancer. 13 years after this discovery made in 1976 almost 50 potential oncogenes have been uncovered. Under normal conditions, these genes are responsible for controlling the cells' growth and development..[247]

[7]. Typical in this respect is the views by Vladimir Dilman. He writes that "...damage to the genetic apparatus, first and foremost DNA, has great significance. Though this phenomenon lies on the basis of mutations, which is a necessary condition for evolutionary variability, it should be restricted as much as possible in the case of the sex cells as well as the somatic ones, because the accumulation of mutations in both the former and latter can lead to death of the cells and to changes in their vital activity, e.g., due to the development of auto-immune lesions."[248] (p.252)

[8]. "Recent research has revealed that children with certain innate developmental disorders have a high risk of developing malignant tumors. For instance, various forms of leukemia are frequently associated with chromosomal syndromes. Leukosis is 10-20 times more like under Down syndrome; Blum syndrome and Fankoni syndrome is combined with acute monocyte and myelosis leukosis in 10% of the cases. Children suffering from aniridia (lack of iris) are 1,000 times more likely to develop a Wilm's tumor [malignant kidney tumor A.K.]. Also linked with this disorder are congenital cataract, concha auriculae, microcephalia manifested in rued head circumference and mental retardation. Is has also been established that

Wilm's tumor and hepatoblastoma are correlated with hemihypertrophy (hypertrophy of half or some part of the body), exophthalmos, macroglossia, and benevolent hamartoma (tuberous sclerosis, heart rhabdomyoma). It has been determined that dysgenesia of gonads is frequently correlated with dysgerminoma as well as with polypoid hamartoma, mucous membrane of the gastro-intestinal, respiratory, and pertaining to urination tracts. Malignant lymphoma are accompanied by developmental disorders of the immune system, growth disorders in long tubular bones, and skin appendixes (hair, nails). However, neuroblastomes are not generally accompanied by growth disorders.

So some tumors in children are correlated with certain growth anomalies while other are not. This attest to the selectivity on the part of these disorders with tumors of specific histogenesis.

All the data linking tumors with growth disorders suggest a common etiology. It is assumed that exposure to exogenous oncogenetic factors short term impact on the fetus results in developmental disorders and long exposure to the development of malignant tumors".[249] (p. 234)

[9]. My metaphorical terms *renaissance* and *renovation* are borrowed from Michail Sergeev's analysis of the development of religion in present-day Russia (source: private discussions).

# CHAPTER 9

# CHARACTERISTICS OF CANCER

## 1. CANCER STAGE BY STAGE

Let us examine the stages of cancer within the framework of the proposed definition of cancer as a pathology of the somatic mechanism of change. Our analysis will parallel previously described somatic mechanism operating in the norm; even the sequence of stages will be preserved.

The first thing to note is that cancer is a *multi-stage* process. It incorporates many diverse features that control the various levels of the genetic code (programs which govern the development of a new organism as well as programs that control the scope and depth of changes in the lower-ranking programs). The changes that ultimately prove to be pathological may be initiated at any stage of the process of somatic change.

Another scenario is that cancer cells mainly interact locally at each stage of development, meaning they interact within the framework of a horizontal mechanism.

So, what is the basic sequence of the malignant process?

1. Changes that transform a cell into a cancer cell may take place at various stages of this process starting from the creation of the organism's architecture during the embryogenesis when HOX genes (see section 2, Chapter 3) are active.[1]

We also observe changes that transform a cell into a cancer cell at various levels of the organism including the genetic level. Genes responsible for these changes may be the same ones that are present in a normal cell. It has been recognized that normal and oncogenes have identical nature.

The sources of change that lead to cancer can be internal to the genome, i.e., driven by its inner mechanism of change (the program which changes the program which changes the program that forms the organism). These internal sources of change are least explored and this whole notion raises skepticism on the part of most biologists. Meanwhile, we know that certain elements of the program for changing the program that shapes an organism are identical for both normal and pathological mechanisms of change. This applies to transposons and certain viruses and perhaps proviruses[250] that tend to induce malignancy. The work of Howard Temin has revealed the role of retroviruses in the malignant process by tracking the DNA environment under which the cancerous retrovirus (RNA segment) is able to infiltrate successfully.[251]

At the chromosome level, changes that lead to cancer reflect chromosome damage or pathological recombination.[252] [2]

At cell level sources of change that induce cancer may not be limited to the genome. Mitochondria and other structures in the cell that contribute to its development may play a role.

At the intercellular level, beside nucleotides, for instance, viruses, cancer can be triggered by oncoproteins.[253]

At the level of the organism cancer is associated with the failure of the immune system to halt the development of the cancer cell.

Causes of cancer external to the organism are many and diverse. They may be rooted in the mechanism of sexual crossing or the environment. Crossing may produce inferior combinations of germ cells meaning one of the cells (perhaps the carrier of the predisposition in latent form) or their combination creates a predisposition to cancer. External sources of cancer believed to exert the greatest impact are carcinogenic substances and high levels of radiation. I would like to reiterate that there is too much emphasis on the link between cancer and various environmental factors at the expense of the less explored dysfunction in the internal mechanism of change.[3]

2. Generally speaking, cancer cells are less differentiated although the spectrum is quite wide.[4] A mature epithelial cell turned malignant reduces its level of differentiation substantially (so called negative differentiation) while immature, meaning less differentiated epithelial cells, can turn cancerous in a more straightforward fashion. Cancer cells generally exhibit a reversion back to lower forms[254] appearing even during reversion to ancient forms[255], for instance from eukaryots to prokaryotes.[256]

3. Having become less differentiated the cancer cell changes its behavior not only because some genes are suppressed, this occurs in normally changing cells as well, but also because some of its genes are damaged by exposure to chemicals or radiation. Cancer cells also behave

differently when new genes in the form of viruses penetrate its genetic structure.

4. As a rule, cancer cells becomes more autonomous in terms of acquiring nutrients and excreting byproducts of their metabolism; they fail to follow normal "input/output" processes regulated by the organism. This drastic change in cancer cells' metabolism is expressed, among many other changes, in the reduced number of receptors linking it to other cells. As opposed to normal cells of complex organisms that consume glucose and oxygen the cancer cells' oxygen supply is upset and a switch is made to a highly simplified form of anaerobic metabolism which only requires glucose from the outside. This characteristic of cancer cells was first pointed out in 1923 by Otto Warburg and became a pivotal element of his general theory of cancer.[257] However this mode of metabolism produces greater amount of lactic acid.[5] In a normal cell operating under anaerobic metabolism lactic acid is eventually discharged and, in fact, it gives back some useful substances in exchange for glucose received from the host. Essentially the cancer cell turns into a criminal. It consumes glucose and excretes waste in the form of lactic acid which the host must somehow utilize [6]; being less differentiated and lacking many receptors the cancer cell is not only oblivious to the needs of the host but is unable to produce many substances needed by the host. In the view of Zinovy Chereisky what makes cancer cells especially dangerous is their pathological metabolic process resulting in excessive concentration of lactic acid coupled with a clogged membrane that prevents its discharge.[258]

5. The threat posed by changes in cells that have turned malignant varies. Cancer cells may result in benevolent tumors. Some tumors are localized to a single spot and stay confined to that organ; there are tumors which move from tissue to tissue within a given organ. These relatively mild forms of cancer interfere with the organism's performance but they are not lethal.

Another feature of cancer cells is piling up of malignant cells that do not exhibit frantic growth.[259] Here, the organ housing the malignant tumor disintegrates.

It would be interesting in this connection to look at tumors from the standpoint of developmental biology and organ deformities. Presumably, tumors represent either an atavism of the mechanism of reproduction via the somatic cells (this incorporates the process of change) or an unregulated embryonic development.

Finally, when the changed cells attain a certain level of maturity they may attempt to leave a given organ, that is begin to metastasize.

6. The cancer cell undergoes change at various stages of its development, including M1 and M2, induced by telomerase activation and resulting telomere renewal.[7]

7. The mechanism of repair of modified DNA segments fails to engage or does so too slowly. "The Beckman team does have a handle on the relationship between slow DNA repair and genetic mutations found in skin cancer."[260] It is to be expected that the repair mechanism does not respond to innovative changes since if it did, it would be like a repairman who, seeing a modernized piece of machinery, disassembles the newly added parts thus reverting to the original state. The dysfunction in the repair mechanism is actually its inability to address harmful changes in the genome which sidetrack the normal course of innovation. This flaw may stem from some defect in the repair mechanism or in the mechanism of change incorporated into a higher-level program.

Another plausible scenario is that the aroused mechanism of change in the cancer cell halts the process of apoptosis thus replacing definite death under apoptosis with possible death resulting from cancer.

8. Growth of the cancer cells is made possible by the activation of telomerase and the resulting change in telomeres (see above paragraph 5).

9. Cancer cells multiply faster than normal cells. There is plenty of literature on the mechanism of rapid growth of cancer cells so I need not dwell on it.

10. Cancer research has been probing into genes responsible for the accelerated rate of cell growth. In the recent years more research has been focused on the reasons for the deactivation of the so called "tumor suppressor genes" that control the process of normal growth.[261] Interesting in this connection is the mutation of the *p16* gene found in many kinds of tumors including skin, urinary bladder, breast, and kidney cancers. The forms of cancer in which the mutated *p16* gene is present exceed considerably the mutations of another recently discovered suppressor gene, *p53* (it was found in rectal and some other types of cancer).[262] The *p53* gene is semifunctional. It is geared toward halting cell division or even directing the cell to self-destruct if its DNA is damaged and the repair mechanism is unable to fix the problem.[263]

11. Cancer cells grow to form tumors which require an internal infrastructure. One of its components is the blood supply system. The same growth/inhibitory factors that contribute to the formation of a new vascular system in normal tissue support the formation of the blood supply system in tumors. What is pathological in this case is that this blood supply system contributes to the destruction of the organism.[264] (p. 124). It is still unclear

whether organisms possess special mechanisms that prevent the formation of a blood supply system around dangerous growths.

12. Most tumors are not localized to a given organ. Cancer cells are capable of transmitting information to cells in other organs through the viruses they produce.[265,266]

Cells contain genes that turn on the mechanism which allows a cancer cell to leave the tissue and invade other organs via the blood/lymph stream.[267]

13. Metastasizing does exhibit certain patterns. Some forms of malignant tumors metastasize very selectively. Metastases are not blown around passively through the blood stream but take root only in receptive organs.[268] Reasons for settling in some tissues and not in others were discussed above in connection with the migration of "innovator" cells. With regards to metastases, it seem they take root in those organs that: 1) consume/supply ingredients to the organ where cancer originated and/or 2) are morphologically connected and/or 3) are formed sequentially as the genetic program unfolds. These considerations are purely speculative since I have no experimental evidence to support these hypotheses regarding the logic of the metastasizing process.

Finally, the presumably iterative nature of the process of coordination of changes induced by the somatic hereditary mechanism would cause metastasis to return to the original organ in the form of cells "enriched" with the newly acquired information or as cells of other organs that have been changed.

14. Another crucial variable is the ability (or rather inability) of the immune system to combat cancer, to check its devastating activity. Presumably there is a certain mechanism that weakens the effectiveness of the immune system, at least in so far as not to impede the emergence of dissident cells, including cancerous ones. Viewed in this light, the problem of combating cancer by strengthening the immune system takes on an added complexity. In the case of AIDS, it is possible to argue that the suppressers of the immune response have gained enough strength to make the organism helpless in resisting various pathogens. A correlation of AIDS and cancer points to a similar progress of immune deficiency with respect to cancer cells allowing them a chance to grow uncontrollably.[8] It cannot be ruled out that ultimately the immune system does not differentiate at all between cancer cells and innovator cells, although it is rather effective in recognizing foreign organisms based not only on the telltale features of their protein coats but also, as we now know, their DNA pattern (so far this feature is found limited to microbes and viruses). These

results are based on the specific features of bacterial DNA and their impact on the immune system.[269]

Whatever the reasons for immune deficiency in combating cancer, many researchers have proposed strengthening the immune system with various drugs. Of speci al interest here is the pharmacological research conducted in the U.S by Steven Rosenberg at the National Cancer Institute and Hilary Koprowski at the Wistar Institute in Philadelphia.

The link between cancer and the immune system calls for further investigation. I have found it beyond me to enter the highly complex field of immunology. But I certainly hope to attempt this incursion in a not too distant future for in this book I have already sown the seeds of this investigation by presenting the analogy between society and the immune system.

To sum up our discussion I would reiterate that cancer represents a systemic disease that embodies the striving by all living creatures for change. The phenomenon involves radical deviants and the inability of the immune system to address them, i.e., to assimilate innovation. In other words, the damage is not caused by the appearance of newly formed radical cells but by the inability of the organism to assimilate them, i.e., to gradually integrate the cells' innovativeness, making it part of the general process of biological development. By analogy with social systems introduced in the previous chapters (hopefully it has helped elucidate the point) political systems incorporate a variety of programs including radical ones. The power of the extremists lies in their ability to propose new ways of development, to create ideals to be aspired toward, even if these ideals are unattainable. The weakness of the extremists is attempting to immediately achieve the set goals, perhaps unattainable altogether, and bypassing all the intermediate stages.[9] The development of a system as a whole requires that no extremist group be allowed to seize power. So the danger stems from the inability of the system to deal with radicals rather than their presence.[10]

General arguments aside, assuming that cancer represents a pathological manifestation of the somatic mechanism of change I have put together in the following table all the alleged stages of development of a normally changing, i.e., innovator cell, contrasting these stages with experimentally confirmed stages of cancer.

TABLE 9.1. Stages governing normal and pathological change in somatic cells.

| Stages of change in the norm | Stages of cancer |
| --- | --- |
| 1. The sources of change range from external factors such as chemicals, radiation, viruses to internal self-inducing processes. | 1. The sources of change range from external factors such as chemicals, radiation, viruses to internal self-inducing processes. |
| 2. It seems that the more pervasive the changes affecting the cell the less differentiated the cell must be in order to free itself of the forces suppressing the development of its diverse genetic capacities. | 2. As a rule cancer cells are less differentiated although the spectrum is quite wide and includes regression to ancient forms. |
| 3. The less differentiated cell alters its performance due to changes within its genetic structure or via the inclusion of new components into its genome. | 3. The main reason for lesser differentiation of the cancer cell is the suppression of some of its genes. Cancer cells also form by incorporating new genetic elements, such as viruses, into their genetic structure. |
| 4. The innovator cell strives for greater autonomy by simplifying its metabolism (anaerobic process supersedes oxygen intake), reducing the number of receptors linking it to other cells, etc. | 4. Typically, the cancer cell becomes more autonomous in its consumption of nutrients and discharge of byproducts, radically simplifying its metabolism, reducing the number of receptors, etc. |
| 5. The scope of changes in the cell is quite wide. Roughly we can distinguish three phases: minor changes, significant changes, and major changes. | 5. The devastation caused by changes in the cell turned cancerous varies from benevolent tumors to metastasis. |
| 6. Changes are implemented by different mechanisms, including telomeres, depending on the stage of cell development. | 6. At the various stages of its development, including M1 and M2, the cancer cell undergoes change induced by the activation of telomerase and the resulting change in telomeres |

7. The repair mechanism in a cell is possibly turned off when it encounters an innovation.

7. In cancer cells the repair mechanism fails to address, or is slow in doing so, the damaged segments of the DNA; it seems that the apoptosis mechanism also fails to perform.

8. Following its newly acquired specialization the cell begins to reproduce owing to telomere renewal.

8. Cancer cell growth is made possible by activation of telomerase and the resulting changes in telomeres.

9. The specialized cell must grow faster in order to affect the development of related cells.

9. Cancer cells also multiply faster than normal cells.

10. The presence of growth accelerating genes suggests the presence of growth-arresting genes.

10. In cancer cells the genetic growth-arresting mechanism malfunctions.

11. The changed cells begin to form new structures that require an infrastructure including the blood supply system.

11. Cancer cells form a malignant tumor with own infrastructure.

12. Methods of transmitting information to other cells include secretion of genetic information as well as migration of the entire cell into other organs. In order for a cell to invade other organs there must be a mechanism of cessation from the living tissue of the host organ.

12. Most cancer cells are not localized to a given organ. They can transfer information to cells of other organs. There are genes in the cell that allow cancer cells to disengage from the host tissue and intrude into other organs through the blood or lymph stream.

13. The settlement pattern of a cell in other organs is probably not random.

13. There is method to the process of metastasizing.

14. In order for an innovator cell to fulfill its role each one must be to some extent "politically" independent, i.e., the immune system should not interfere too much with its development. Possibly, this task is fulfilled by special regulatory genes.

14. Disparity between the potency of the cancer cell and the immune system allows the former to develop. Purportedly AIDS involves intensified activity by the genes that block the immune system thus making an organism defenseless against all kinds of deviants.

Assuming the overall approach makes sense, I would conclude that the same process of change can be directed at creating new more perfect organisms as well as at destroying the existing ones. The implications in terms of program hierarchy (see Introduction) are the following: the first-level genetic programs (that change the zero-level program that forms an organism) for creating innovation are quite identical to the ones that destroy by means of cancer what is already there. The fact that normal genes and oncogenes responsible for change (that is belonging to the second-level program) are quite similar indirectly corroborates this speculative idea that has become commonplace in the biological establishment. Nonetheless, the mechanism underlying the interaction at various stages between the genes belonging to the first-level program and genes of the zero-level program is still obscure.[11]

Perhaps the difference in the genes responsible for the development of cells bearing constructive innovations or malignancies resides in the second-level program, the program which changes the first level program. The second level program is the main sanctuary for genes that cause excessive behavior on the part of cancer cells both in isolated instances of change (excessive loss of differentiation, rapid growth, propensity to invade other organs) as well as their general aggressiveness.

## 2. CANCER AS A PATHOLOGY OF THE MECHANISM OF CHANGE: SOME IMPLICATIONS OF THIS APPROACH

### 2.1. Cancer and Age

Strictly speaking cancer should be associated with the state of an organism rather than its age. Each particular case is different but a sufficiently large set of individuals will exhibit statistical patterns since age is strongly correlated with the state of an organism.[12]

Disorders in the mechanism of change vary depending on the stage of life.

During the embryonic stage when the focus is on the formation of an organism and there is much room for variation on organs, pathological changes in the genetic code are manifest, primarily in deformed fetuses, still-born babies, etc., rather than cancer.

During the stage of development but prior to puberty when growth is the characteristic feature, negative deviations in the mechanism of change lead to deformities, but in much smaller quantities than at the previous stage.

The frequency of cancer may rise as the organism matures. This may be correlated with increased activity of the mechanism of change (under a very strong assumption that changes that have been initiated may be slowly getting ready to transform to the progeny via the germ cells). This process could be more pronounced in females; these observations agree with the fact that in girls cancer occurs at an earlier age than in boys.

At the stage of sexual maturity but prior to aging cancer is on the rise but disorders in the mechanism of change at this stage of one's life are not that common since both the mechanism of change and the immune system are still rather stable.

Old age predisposes toward break-downs of the still active mechanism of change associated with the organism's reproductive capacity as well as weakness of the immune system. As Vladimir Dilman mentioned, "... it has been statistically noted that frequently the reproductive function has been switched off later than usual in women who fall ill with breast cancer after menopause."[270] (pp.120-121)

In deep old age when sexual activity and its companion the mechanism of change decline rapidly, the frequency of cancer declines in spite of decreased robustness of the immune system. For instance, we observe a decrease in such widespread form of cancer as female breast cancer.

Frequency of cancer is strongly correlated with the aforementioned stages of the organism's development, and the highest rate of cancer cases corresponds to old, but not deep age. From 1972 till 1990 4,894 cases of death were diagnosed in the Tokyo Metropolitan Geriatric Hospital. A well defined tendency for the share of deaths from cancer to go down with age was found: among 60 year olds the percent of cancer patients was 50%; among 70 year olds - 47.9%, 80 year olds -43.2; and 39% among ninety years olds and older.[271].

I would say that my general hypothesis on cancer agrees with many empirical observations on the behavior of cancer cells under one very important qualification, that certain somatic changes can transfer to germ cells.

## 2.2. Cancer and the Sexes

Considerations expounded in the previous chapter concerning the role of the sexes raise interesting questions regarding cancer of male and female reproductive systems.

TABLE 9.2. Cancer frequency in male and female reproductive systems (US, 1991).[272]

| Organs for | Quantity | Sex | |
|---|---|---|---|
| | | Males | Females |
| Production of germs | 26,800 | 6,100 | 20,700 |
| Ovary | 20,700 | | 20,700 |
| Testis | 6,100 | 6,100 | |
| Delivery of germs and development of fetus | 350,100 | 124,100 | 226,000 |
| Uterus | 46,000 | | 46,000 |
| Breast | 175,900 | 900 | 175,000 |
| Prostate | 122,000 | 122,000 | |
| Others | 6,200 | 1,200 | 5,000 |
| Total | 376,900 | 130,200 | 246,700 |

As expected, changes in the female system lead to more cancers if only because of the greater complexity of the female reproductive system. Statistics shows that in 1991 in the US for 247,000 cases of cancer of female reproductive organs there were about 130,000 cases of cancer of the male reproductive organs.

My hypothesis seems to point to other differences in the cancer of male and female reproductive organs.

Let us examine the subsystem responsible for the production of germ cells.[273] For instance, ovarian cancer is significantly more frequent than cancer of its functional counterpart, the male testis. The chief cause for this discrepancy is that the somatic cells that have undergone change in other organs have a relatively easy time invading the ovaries, while penetrating the testis is practically impossible. Of course there may be many other reasons underlying the frequency gap between cancer of male and female organs and it is rather difficult to isolate and assign weight to the barrier factor, but it might prove to be a viable working hypothesis.

Statistics also shows that in 1991 there were 21,000 new cases of ovarian cancer and 6,000 new cases of cancer of the testis. If we take into account the potential destructive power of the two types of cancer the difference becomes much more pronounced. For instance, that same year ovarian cancer resulted in 12,000 deaths and cancer of the testis caused about 400 deaths, a 30-fold difference.

Of course, ovarian cancer should not be equated with changes in the ova. By invading the ovaries cancer cells may change rather than destroy them, thus eventually (maybe after one generation) changing the ova. Cancer cells can invade the ova directly, thus destroying them. It would be interesting to explore how cancer cells alter the genetic structure of the ova. A similar investigation of the possible changes in the sperm could be conducted in those rare cases when cancer cells penetrate the testis.[274,275]

Table 9.2 seems to suggest that most types of cancer of the reproductive system strike the subsystem responsible for germ-cell delivery and fetus development. This is true for both men and women. However, male and female organs comprising this subsystem cannot be compared. Although male and female breasts are somewhat similar, they operate on rather different planes. So the observed quantitative gap between male and female breast cancer is not unreasonable.

The male organ most frequently afflicted by cancer (more than 90% of all cancer of the male reproductive system) is the prostate gland. In terms of numbers prostate cancer (120,000 cases) is of the same order of magnitude as female breast cancer (more than 175,000). I would like to show that these two types of cancer are not completely unlike.

We observe that the great majority of all cancers of the reproductive system (about 93%) afflicts the subsystem comprised of organs that accommodate germ cell movement as well as fetus development in females (including feeding the infant). I noted above that breast cancer prevails in women and prostate cancer in men. I would venture to say that these particular organs are most sensitive to the organism's response to changes in the environment and must implement changes in the respective organs that produce such ingredients as milk or fluid that protects sperm from its antagonist, the urine during the sperm's movement through the urethra. Moreover, considering the importance of these changes for the offspring, it is essential that these changes be transmitted quickly, perhaps even to the germ cells (directly or through other organs). This could explain why metastasizing starts early in breast cancer even when the tumor is relatively small. It would be interesting to compare breast cancer and ovarian cancer from this point of view.

### 2.3. Why are Not All Organs or All Organisms Susceptible to Cancer?

Statistics reveals that cancer does not strike all organs with the same frequency and, in fact, it varies by country and over time in the same country.

Naturally, external conditions, especially such factors as pollution, smoking habits, foods, etc. can trigger cancer.

All these geographic and dynamic differences not withstanding cancer exhibits certain universal frequency patterns. I would like to take an extreme case of certain cancer-free organs although adjacent organs that may be similar in tissue structure are stricken. For instance, cancer of the trachea is non-existent while cancer of the adjacent bronchii, i.e., lung cancer, is widespread. Another example is the duodenum. It is resistant to cancer that does invade the adjacent stomach and esophagus. If cancer starts in the stomach it moves upwards and if it goes downward it terminates at the duodenum. Cancer is also possible at the junction of the liver excretions with the duodenum.

These examples suggest the following very speculative conjecture: the less change experienced by a given organ over the course of evolution, the less prone it is to cancer. The notion that different organs have evolved differently is more or less an established fact.[276] In any case, this correlation between the changeability of an organ and cancer distribution by organs merits an investigation.

At the phenomenological level, if we consider an organism as a whole, biologists observed a certain correlation between organisms' changeability and the respective frequency of cancer. As Meckler points out, the tempo of evolutionary change in a given taxonomic class is correlated with the frequency of tumors it experiences.[277] For instance, marmoset monkeys evolved much faster than baboons and the former experience spontaneous tumors 4-6 times as often as the latter. Tumor frequency among rodents has the following distribution, in decreasing order: most frequent among mice, then rats, rabbits, and finally guinea-pigs. This distribution is correlated with the pace of evolution of these animals. Cartilagenous fishes are the slowest among vertebrates to undergo evolutionary change.

Correlation between the pace of evolution of the given species and tumor frequency is no proof that cancer is a manifestation of an impaired mechanism of change, but at least the two phenomena show positive correlation. It would be fascinating to study this link at the structural level to test whether this correlation underlies a true cause and effect relationship.

Especially interesting in this connection are certain species of much studied fish that have changed relatively little over time and experience very little cancer. I mean the sharks.[278]

We know that sharks have not changed much over the last 400 million years and cancer among sharks is practically unknown. Since shark harvest is rather large - about 5-7 million a year, scholars had an opportunity to verify these claims empirically. In fact, for many years researchers placed sharks in pools containing carcinogenic substances but tumors failed to materialize.

The reason for the sharks' resistance to cancer are unclear. But three factors do enter into the explanation.

The first is the cartilage rather than the bone skeleton of the shark. Its cartilage contains six or seven proteins that can prevent the formation of blood vessels (angiogenesis) to the organ (the cartilage itself needs no blood supply). In moving through the blood stream cancer cells prefer to move through new vessels, as they do not fare as well through the old vessels, and their growth requires an additional blood supply. Therefore, tumors frequently surround themselves with their own blood supply system.

The capacity of cartilage, shark cartilage in particular, to block the development of blood vessels, and in fact destroy blood vessels thus depriving the tumor of the blood supply and waste disposal, stirred some interest in using cartilage to make drugs. Its primary use is for cancer treatment, but it could also prove effective in helping treat arthritis, psoriasis, as well as some other ailments.

Quite possibly compounds found in cartilage affect the immune system: in any case cartilage-based drugs have proved effective in treating all kinds of inflammatory processes.

The second reason for the shark's "untouchability" is their extremely powerful immune system. Their wounds heal quickly and sharks are basically free of infections. Antibodies found in the sharks' blood combat bacterial and virus infections as well as protect sharks from all kinds of harmful chemicals that are lethal to many other kinds of mammals. Unlike the human immune system which is normally passive the sharks' immune system is constantly circulating ready to attack.

The third reason is the sharks' unusual capacity to prevent the activation of aflatoxin $B_1$ that stimulates cancer. Aflatoxin itself is a kind of forerunner to carcinogenic substances. When aflatoxin is excited it links DNA with these substances and the resulting compound may attack genes that would otherwise prevent the expression of cancer by cells.

It seems to me that all these safeguards that protect sharks from cancer have to do mainly with a very dormant internal mechanism of change combined with a highly stable and powerful immune system that can prevent or eliminate any change in the shark's organism.

Encouraged by the infrequency of cancer among sharks a number of scholars have expressed optimism regarding the use of shark cartilage in treating cancer. Their enthusiasm was not welcomed by biologists at large.

In October 1993 Tim Beardsley published a brief article "Sharks Do Get Cancer. Cartilage cure relies on wishful thinking."[279] The title itself implies that sharks' immunity to cancer is a pseudoscientific myth and cancer treatment with shark cartilage relies on wishful rather than factual data. The article dethrones the aforementioned book by W. Lane and L. Comac whose title Sharks Don't Get Cancer suggests just that and advocates cancer treatment methods using shark cartilage.

Beardsley referred to data provided by reputable scientific organizations that shows 20 registered cases of cancer in sharks, including shark cartilage. The authors of the targeted book do not deny that sharks get cancer, but it is very rare. The fact that 20 cases have been documented does not refute Lane's and Comac's basic contention. A more serious argument presented by Beardsley quotes John Harshbarger, director of the registry of tumors in lower animals at the Smithsonian Institution in Washington, D.C.: "data do not exist to determine whether sharks get cancer more or less often than do other creatures." However, evidence for the lower frequency of cancer among aquatic animals as well as the statistics gathered by Land and Camac does indeed support the claim that sharks are less predisposed toward cancer than other creatures.

Perhaps, the authors' enthusiasm in advocating cancer treatment with shark cartilage is indeed excessive. But Beardsley himself acknowledges that under certain conditions shark cartilage can be beneficial in treating cancer.

Beardsley also notes that The National Cancer Institute tested a number of venues of treating cancer using shark cartilage, but the results were less than promising. At about the same time that Beardsley's article appeared in <u>Scientific American,</u> <u>Wall Street Journal</u> published a report by Chrisn Frampton, "Alternative Medicine to Treat Cancer Undergoes Mainstream Study by NIH" (National Cancer Institute is part of National Institutes of Health). According to the report NIH will conduct research into the use of shark cartilage in cancer treatment. Evidently the data cited by Beardsley has not deterred NIH from conducting further research.

My critical attitude toward Beardsley's article is not meant to defend the proponents of shark cartilage treatment. Generally speaking, I am not a big fan of alternative medicine if only because the scope of its application is not well defined. But what are we to do in the face of this devastating disease? We lack methods of treatment based on profound understanding of the malignant process as well as cures effective as far as practical treatment is concerned. Under the circumstance, alternative solutions that seem promising, or at least not harmful, should not be ignored!

## 3. QUASI HYPOTHESES STEMMING FROM THE MATERIAL PRESENTED IN THIS CHAPTER

I have formulated a number of quasi hypotheses amenable to experimental verification based on my definition of cancer as a radical pathological attempt to restructure an organism by means of the somatic mechanism of change.

These hypotheses that I have tried to corroborate throughout the present work are presented below:

1. Cancer is a disorder at the second level of a hierarchically organized internal mechanism of change (the zero level genetic program controls development, the first level program changes the zero-level one, and so on).

2. The frequency of cancer is directly proportional to the species' predisposition toward change and, inversely, to the defense afforded by the immune system.

3. Quite possibly, under the assumption that certain changes in somatic cells do transport to germ cells, cancer occurs most frequently during the period when the organism begins to age, but prior to deep old age. The reproductive system through which changes are passed on and which supports the mechanism of change begins to grow weak, but does not fade away completely as does the mechanism of change. The immune

system which prevents the formation of pathological changes suffers a similar decline.

4. The frequency of cancer of specific organs is directly proportional to the organ's predisposition to change and inversely proportional to its defense by the immune system.

5. In one particular case of male and female reproductive organs that produce germ cells, the difference in the respective frequency of cancer stems from the functional role of the two sexes (male characteristics are geared toward assimilation of changing environmental conditions while the female system is structured to accommodate major changes in the organism).

6. Metastasis take root in organs that are "suppliers/consumers" of the organ where the malignancy originates. The extent of infiltration depends not so much on the capacity of the vascular/lymph system to carry the cells but the scope of change demanded of the "consumer/supplier" organ. Another possibility is that organs susceptible to metastasis are morphologically linked. Finally, cancer may strike organs that are formed according to the sequential logic of the genetic program as it unfolds.

7. The process of change via the somatic cells incorporates a return of the messenger cell to its host organ; this holds for cancer cells as well. Therefore, even if cancer has disappeared from a particular organ, it can relapse because the emigrant cancer cell has come back rather than because new cancer has flared up, especially after the organ has undergone radiation, chemotherapy, or surgical treatment.

8. Assuming cancer cells are of a dissident type, rather than destroy these cells treatment should aim to limit the scope of their activity, possibly by isolating them temporarily. Under extreme circumstances when the danger posed by cancer cells is lethal and their profusion cannot be curbed, they must be removed from the organism.

## NOTES TO CHAPTER 9

[1]. "We have studied HOX gene expression in several human tissues and organs as well as in their neoplastic counterparts. We have observed (a) characteristic patterns of HOX gene expression for each normal solid organ analyzed, (b) altered HOX gene expression in kidney and colon cancer, (c) a correlation between HOX gene expression and different histological types of primary small cell lung cancer (SCLC) and (d) marked alterations of HOX gene expression among primary and metastatic SCLC variant types. Furthermore, we have shown that differential patterns of HOX gene expression correlate with the adhesion profile (VLA-2, VLA-

5, VLA-6 and ICAM-1) and N-RAS mutation in clonal melanoma populations isolated from a single human melanoma metastasis. This suggests that HOX genes act as a network of transcriptional regulators involved in the processes of cell to cell communication during normal morphogenesis, the alteration of which may contribute to the evolution of cancer."[280] pp. 38-49.

[2]. N.Vorontsov gives the following description of cancer at the chromosome level:
"In case of pathology, the rodents die as a result of papillomatosis disease before one year of age having bred several litters. Fixation of the macromutation of papillary structure of stomach as a specific character seems to be related to a shift of macrovilli formation to the early stages of ontogenesis, while the lethal effect of the neoplasm is shifted to the postreproductional period. Evidently, in this case the macromutation fixation can be accompanied by a selection of variants submitted by heterochrony."[281] pp.53-54.
See also O. Oliveros, et.al. [282]

[3]. In children cancer is not strongly correlated with external factors. In adults many carcinogenic substances enter the system through breathing. At the same time cancer of nasopharynx is rare and cancer of the trachea is altogether unknown while cancer of the lungs is widespread.

[4]. If decreased differentiation merely involved cell *degeneration* to an earlier stage and was not accompanied by cell *rebirth* the natural approach to treating cancer at that particular stage would be to reinstitute the normal level of differentiation. The problem is how. Techniques used to treat trophic ulcer which involves disintegration of the tissue with no rebirth provide some clues. One method helps the tissue itself to recuperate. It was assumed that tissue "management" on the part of the nervous system appeared at a later stage of evolution and was thus limited to the more developed groups of tissues. These considerations suggested to the following methods of treatment. Roughly speaking, the links between the disintegrated part of the tissue and the nervous system were severed by novocain blockage and tissue feeding was done by a compress. When the process of tissue recovery warranted reconnection with the nervous system the Novocain blockage was suspended.
I realize the simplicity of this analogy for devising methods for treating cancer.

[5]. "A frequent characteristic of many malignant tumours is an increase in anaerobicglycolysis, that is the conversion of glucose to lactate, when compared to normal tissues."[283]

[6]. The quoted article reads: "In some ways many malignant tumours behave much like parasites, drawing upon the host for nutrients such as glucose and returning waste productssuch as lactate to the host for recycling or disposal." [284]
I prefer to call these cells criminals rather than parasites. Parasites do not generally ruin the host or force the host to process their waste (oftentimes parasites engage in beneficial activity). Criminals, on the other hand, not only seize things from the master and ruin him but may also force him to work for them or even murder him.

[7]. "The speculation that telomerase may play a role in human cancer is not new and has been discussed in a variety of contexts. However, a study by Counter, *et all* demonstrates that telomerase is activated in ovarian carcinoma. Their data indicate

that expression telomerase and the resulting stabilization of telomeres may be important for the expansion of a human tumor."[285]

[8]. Higher correlation between AIDS and cancer was discovered in patients needing organ transplant. "AIDS patients and children with inherited immune deficiencies also have higher than expected rates of certain tumors, including some of the same cancers found in transplant patients."[286]

A particularly striking correlation between AIDS and Karposi's Sarcoma was discovered. "Approximately 15% of AIDS patients have the sarcoma, which makes it 20,000 times more common in the AIDS population than in the population at large."[287]

[9]. Assuming the cell with its "desire" to fulfill its biological potential forms the "epicenter" of a multi-cell organism there are a number of options for cells to achieve this end. The option selected (manifest primarily in the "urgency" of cell action) depends on the type of cell. An analogy with society should clarify the point.

Different social systems may eventually arrive at the same result aspired toward by many humanists: to ensure that each individual is able to fulfill his/her potential. Of course it is difficult to determine the minimum amount of money (subsistence level) needed to implement this goal. It varies in different countries and at different periods. But whatever this minimum level is history has presented us with many alternatives. One quality that distinguishes different options is the rate at which some minimum is to be achieved. This feature reflects the difference between communists, social democrats, and the bourgeoisie.

Communists content that the production capacities are sufficient for this goal to be attained *immediately* provided the chaos of the marketplace is conquered in the revolutionary manner, public property, plan, and income redistribution instituted.

Social democrats claim the goal must be achieved step by step. It should be actively pursued by the government through steep progressive taxes, nationalization, etc. Communists want to attain the goal quickly and by any means whatsoever, while for the social democrats the process itself becomes the crucial issue.

Competition based structure of a bourgeois society with the various political forces provides every human being an opportunity to realize his/her potential no less successful than in countries with social democrats in power (not to speak of the communists).

[10]. West German President, Richard von Weitzaker, in a speech on May 24, 1989, commemorating the fortieth anniversary of the establishment of the Federal Republic, said that in his country "The present generation of policy makers is made up of people who learned from history. This generation understands that the tragedy of Weimar was not that it produced too many extremists too fast, but that for far too long it had too few democrats."[288] p.153.

[11]. "Some of the cellular changes underlying the presentation of cancer in a patient can already be understood in terms of mutations affecting specific gene functions. So far, only a few of the mutated genes responsible for carcinogenesis have been identified and these are chiefly involved in deregulation of cell growth rather than with the processes of invasion and metastasis."[289]

[12]. An analogous situation arises with social security received upon reaching a certain age. All the practical complications aside, pension should be awarded based on one's condition, meaning the individual's ability to continue to work. However the problems involved in implementing this individualistic approach preclude its enactment.

# CONCLUSION

## 1. INTRODUCTION TO CONCLUSION

The crux of the present work is the inclusion of the "sector of change" with its diverse methods and mechanisms, including random mutations, as an endogenous part of the biological system, similar to the R&D sphere organically intertwined as an endogenous part of the economy.

The reason I dwell so much on this notion of change is because this phenomenon defines life. Other conditions being equal, the waning of the organism's capacity for change may result in painless death. Recent discoveries concerning the possible link between telomeres and the process of change seem to validate this statement. It goes without saying that other factors may cause the death of an organism long before its capacity for change is exhausted. Pathological disorders in the mechanism of change are one cause of death, especially when they assume such devastating form as cancer (just as malfunctions in the R&D mechanism can lead to the disintegration of a social system).

The spotlight in the present work was on the somatic mechanism of change. The fact that change in individual somatic cells takes place is not contested. What is not clear is whether or not there is a *multi-stage* process of change based on somatic cells. General evolutionary considerations suggest an affirmative answer since the phenomenon of biological change existed prior to the appearance of specialized germ cells. Even if somatic change does take place in developed organisms, it may represent some archaic mechanism that may or may not interact with the germ cells that pass on hereditary information.

In my opinion, biological science is just beginning to probe into the normal process of organism restructuring through the internal somatic mechanism of change. Research on the subject could enrich our understanding of disorders incorporated in the more general process of biological change. Furthermore, the paradigm that regards a disease as a break-down of some otherwise normal mechanism points to new ways of reconstructing the normal mechanism through these manifest deviations.

All these considerations went into my definition of cancer as an *extreme form of a pathological attempt to reconstruct an organism via the somatic mechanism of change.*

My ruminations into the mechanism of somatic change, its pathology, and some related issues are encapsulated in the following quasi-hypotheses".

## 1.1. Quasi -Hypotheses Concerning the Normal Mechanisms of Change

1. The cell's genetic system possesses a hierarchically organized internal mechanism of change. The program that codes for organism development is denoted as the zero level program; the program that changes it is the first level program and the program that changes the first level program is called the second level program.

2. Taking into account the intricate nature of the generative system and the resources required to sustain it (see, for example, the C-value paradox in Chapter 3) the organism's development raises the problem of resource allocation, creating and nourishing a dynamic generative system on the one hand, and faster cell growth and multiplication governed by a routine generative program on the other.

3. It is quite plausible that the genetic system incorporates the so called tunnel process. In other words, change is initiated both at the end, i.e., induced by the environment which can directly affect the zero-level program with chemicals, radiation, etc., as well as at the beginning, i.e., internal changes in the second or first level programs.

4. With changes initiated at the beginning there ought to be structures in the genome that minimize those genetic combinations that lead nowhere or to a dead end.

5. The development of specialized germ cells, prior to sexually differentiated cells, and the germatic mechanism of change that succeeded the somatic one (on the evolutionary time scale) was dictated by the advantages of implementing the process of change within one type of cells – in one place where it is much easier and faster to coordinate all the changes.

# CONCLUSION

6. While the somatic mechanism of change plays a secondary role perhaps it continues to fulfill the following functions: a) it complements the germ mechanism in situations where relatively minor and slow changes are required by the organism; b) it fulfills specific functions not covered by the germatic mechanism; c) it acts as a back-up ensuring that such a critical evolutionary task as change is not neglected; d) finally, it may be an anachronism.

7. Minor changes unfold primarily from the end while major changes originate at the beginning.

8. Sexes can be defined based on the role of the organisms directly partaking in the act of crossing and change implementation, unlike classification based on the inherited unchangeable function in a group, such as workers and soldiers among insects.

9. In principle, crossing could involve more than two sexes. By analogy with the separation of power in the realm of social systems, there could be at least three sexes: the first is a counterpart to the legislative branch, the second to the executive branch, and the third to the judicial branch. The last sex would oversee that the programs introduced by the other two sexes agree with the fundamental program of development. This *prevents* the formation of organisms that fail to conform to the basic laws of development.

If one were to seek out the third sex in nature, one should probably do so among developed organisms (such as insects) where the population density is rather high making it easier to meet representatives of all sexes needed for reproduction.

If the third sex is not found in nature, one could set up computer simulation of the evolutionary process under the assumption of trisexual reproduction and see what kind of results are obtained.

10. It is quite possible that under mating the male expresses primarily the end phase of the process of development (i.e., he is the vehicle of environment-induced changes). The female sex is primarily involved with the beginning phases, meaning profound changes in the structure of the organism. This does not prevent each sex from engaging in the functions specialized in by the other sex.

Perhaps, the above hypothesis could be rephrased: "Why is it that among developed animals, the testicles are outside the body and the ovaries are hidden deep inside under the skin?" The answer proposed in the books does yield to experimental verification.

11. In trying to reconstruct the normal mechanism of somatic change through its pathological manifestation, in this case metastasis, one

could hypothesize migration of normal somatic cells (perhaps, some emigrant cells even return to the original organ). Moreover, cell migration, apart from embryonic cells, blood and lymphatic cells, is not chaotic, meaning it is governed by the logic of the process of change. The fact that even sporadic migration of normal somatic cells has not been documented does not mean it is non-existent. A number of biologists have confirmed that the hypothesis of somatic cell migration does yield to experimental verification.

12. Since activation of the telomerase involves all cells undergoing change, natural death or fading away of an organism could be due to the termination of the process of change. Some scholars have entertained the idea of extending the organisms' life span by direct manipulation of the telomerase. It seems that this method which leaves out the mechanism that excites telomerase under normal conditions may upset the balance of related physiological processes.

## 1.2. Quasi -Hypotheses Concerning Cancer

1. Cancer is a disorder at the second level of a hierarchically organized internal mechanism of change (the zero level genetic program controls development, the first level program changes the zero-level one, and so on).

2. The frequency of cancer is directly proportional to the species' predisposition toward change and inversely to the defense afforded by the immune system.

3. Making one strong assumption that certain types of somatic change do transfer to germ cells, it seems cancer occurs most frequently during the period when the organism begins to age, but prior to deep old age. The reproductive system through which changes are passed on and which triggers the mechanism of change begins to grow weak but does not fade away completely as does the mechanism of change. The immune system which ought to prevent the formation of pathological changes suffers a similar decline.

4. The frequency of cancer of specific organs is directly proportional to the organ's predisposition to change and inversely proportional to its defense by the immune system.

5. In one particular case of male and female reproductive organs that produce germ cells the difference in the respective frequency of cancer stems from the functional role of the two sexes. Male characteristics are

geared toward assimilation of changing environmental conditions while the female system is structured to accommodate major changes in the organism.

6. It is quite plausible that the powerful barriers preventing cancer cells from penetrating the testes and a lack of such protection in the ovaries attest to the following phenomenon. The anachronism of somatic change is more pronounced in the females since profound changes initiated at the beginning require a more thorough testing of the parts, at least at the informational level, and their subsequent integration - the task that is better suited for the somatic mechanism.

The difference between the somatic mechanism of change in males and females is correlated with the following puzzling fact: there exist powerful barriers in the way of cancer cells penetrating the scrotum but none for the ovaries.

7. Metastasis take root in organs that are "suppliers/consumers" of the organ in which the malignancy originates. The extent of infiltration depends not so much on the capacity of the vascular/lymphatic system to carry the cells but the scope of change demanded of the "consumer/supplier" organ. Another possibility is that organs susceptible to metastasis are morphologically related. Finally cancer may strike organs that are formed according to the sequential logic of the genetic program.

8. Assuming cancer cells are of the dissident type, rather than destroy them treatment should aim to limit the scope of their activity, possibly by isolating them temporarily. Under extreme circumstances when the danger posed by cancer cells is lethal and their proliferation cannot be curbed, they must be removed from the organism.

To summarize the ideas expounded in the book, I would say that the book paints an apparently contradictory picture of my theory of the mechanism of biological change, bothnormal and pathological. In other words, ideas submitted for readers' judgment conflict with the prevailing neo-Darwinist precepts (mutations generated by the minor random nature of changes in the DNA program bombarded by external factors and tested for viability through natural selection. I do not discard the neo-Darwinist conception. It is not in my character to reject outright all the previous scientific achievements. I believe, following Bakhtin's footsteps,[290] that new theories do not annul the old but merely limit the scope of their applicability.[1] I always try to stipulate the precise conditions under which old theories are valid. However, even mathematics that has excelled at linking the final conclusions with all the initial conditions is full of pitfalls.[2] Of course, I am unable to specify the conditions under which the neo-Darwinist views on the mechanism of change are true. Based on some general considerations and by analogy with economics I would say that these conditions reflect the case of organism modification. On the other hand, my approach underscores the process of change related to the emergence of new species, orders, and classes.[3]

I would also like to note the concept presented in the book makes no claim for immediate acceptance on the part of biologists or for a prominent place in the textbooks. My goal is much more modest, namely for the book to become part of the library of ideas regarding the evolution of the biological system. This too imposes certain conditions that must be met in the book. They are that the ideas advanced should not be generally known, obvious, or outright wrong.

Some people I talked to disagree, maintaining that the necessary condition for an idea which aspires to scientific validity to be admitted into the library of ideas is that it be accompanied by hypothesis which yield to experimental verification. This approach to science is the cornerstone of the Anglo-Saxon tradition. The great advantage of this empirical tradition is that it prevents idle and unconfirmed scientific speculation. Its drawback is that peculiar scorn for vague ideas that are not yet ripe for direct experimental testing. I believe this attitude is rooted in the notion that new ideas are easy to concoct and that any scholar can come up with new ideas at will. Or perhaps, it is held that any new idea needs to be elaborated and this requires money which, in turn, requires that the proposed theory be corroborated, at least to some extent; and if the theory advances no experiment-amenable hypothesis, foundations will not even consider the proposal... and so on and so forth.

Whatever the reasons behind this skepticism toward new ideas, I believe the Anglo-Saxon tradition does have its Achilles' heel. The scholarly tradition of continental Europe is different. It certainly holds ideas amenable to experimental verification (not to speak of those corroborated by experiment) in high esteem. However, the continental tradition shows appreciation for the initial, oftentimes feeble steps in the development of new ideas even if there is no clear-cut method of experimental testing.

The reader might well guess that I am a disciple of the continental European tradition [4], at least as far as keeping the ideas from sinking into oblivion. Nevertheless, I did formulate a number of preliminary hypotheses which do yield to experimental testing. The three aforementioned prerequisites for preserving the idea are certainly insufficient for its positive assessment or for initiating research. Practical implementation stipulates many other conditions, not the least of which is finding "devotees" willing to undertake research based on this idea.

I would again remind the reader that my only claim is for the ideas expressed here to be included in the library of ideas. It is very difficult to judge an idea at an early stage and only time will tell whether or not my ideas will attract readership.

It is very tempting to try to assess the importance of these initial stages in terms of investing in subsequent research. However, it is very difficult, if not impossible to *objectively* evaluate innovative theories.

Nevertheless, it has been tried. Perhaps, the most famous attempt is the Science Citation Index devised by Eugene Garfield in 1963. The crux of his recipe was to measure the publication's worth by the number of times it has been cited. This method was widely accepted both among scientists as well as agencies which allocate research money. This formula is unlikely to produce many admirers for my ideas. However, I am optimistic because the above method has recently been subjected to some harsh criticism. Take, for instance, an article which appeared in Science. It quotes Garry Schuster, the head of the University of Illinois chemistry department, as saying:

> "You have to wonder what kinds of papers get the most citations and what that really means. Papers that report useful techniques, for example, will be cited more often than papers that present a new concept, particularly a very new one and one that not many people in a field have thought about. And then review articles in a field can garner a lot of citations and really make no original contribution to the science."[291]

Therefore, the important thing in assessing innovative ideas is not to try to evaluate them objectively, but to create conditions for their preservation. And in case there are people interested in these ideas, these people will have a chance to pursue them based on their own subjective judgment.

Naturally, this final conclusion is an attempt to justify my ambitious incursion into biology, and some of the most intricate and fundamental biological phenomena, the mechanism of evolutionary change and (!) cancer. Moreover, I have made clear my intention to link these two phenomena by viewing cancer as a pathology of the mechanism of evolutionary change.

However, my hope is to find an appreciative reader who will discover some useful information and will be forgiving toward my ambitions.

## NOTES TO CONCLUSION

[1]. "The rise of the computing's new order will be anything but painless. Since mid-1950's the computing world has been overthrown twice: minicomputers broke the monopoly of mainframes in the late 1960's and early 70's, and in the mid-70's personal computers began edging minicomputers out of the spotlight. Each time, the older machine survived, but it no longer ruled."[292]

[2]. My now deceased friend Boris Moishezon told me of certain pitfalls befalling algebraic geometry. At the end of the 19th century a group of Italian geometers proved a number of important theorems. When their proofs were reevaluated in the middle of this century based on the new developments in algebraic geometry many proofs were found faulty in the sense of the final conclusion not being rigidly derived from the initial assumptions. Some of the theorems were salvaged by stipulating the precise conditions under which they are true.

[3]. In a sense, my views recall Løvtrup's[293] concept that new species emerge as a result of macromutations that are independent of other mutations.

[4]. In the extreme case, the dream of speculative-writers is to write in the spirit of socialist realism. I would like to present one real-life example so that an uninitiated reader might understand the essential nature of this style.
In the early 1930s a Soviet journalist, Elena Mikulina,[294] published a brochure about the socialist competition. Stalin wrote an introduction to this brochure because he exalted socialist competition as an antidote to its capitalist counterpart. Mikulina's brochure was written in the notorious tradition of socialist realism, i.e., the material was largely contrived while claiming to describe reality - real life socialist competition. At that time Stalin had a firm, but not an absolute grip on power. So, it was still possible for one Russova, who was unable to criticize Stalin directly for supporting such a superficial article with contrived facts, to do so indirectly by accusing Mikulina "of misleading comrade Stalin". Wishing to shield comrade Stalin from unworthy publications, Russova suggested that this brochure be withdrawn from the shelves. A well-known communist at the time, Felix Kohn, who was the head of the Union Radio Committee and deputy chief of the International Control Commission gave Russova's review to Stalin. In his reply to Kohn Stalin wrote that he realizes that "There is no spinner by the name of Bardin in this world and there is no a spinning factory in Zariadie". However, he (Stalin) still thinks that Mikulina's brochure holds merit because it "**popularizes** the idea of competition and **instills** in the reader the spirit of competition. That is what really counts, not some minor misconceptions."[295]

# REFERENCES

1. D.COOPER, M.KRAWCZAK, Human Gene Mutation (BIOS Scientific Publishers, Oxford, 1993).
2. B.ALBERTS, D.BRAY, J.LEWIS, M.RAFF, K.ROBERTS, and J.WATSON, in Molecular Biology of the Cell (Garland Publ., New York,1989).
3. L.WHYTE, Internal Factors in Evolution (Tavistock Publ., London, 1965).
4. H.BERGSON, Creative Evolution (University Press of America), 1983).
5. The Crisis in Modernism: Bergson and the vitalist controversy, edited by F.Burwick and P.Douglass (Cambridge University Press, New York,1992).
6. H.FREYHOFER, The Vitalism of Hans Driesch : the Success and Decline of a Scientific Theory (Peter Lang, Frankfurt an Main, 1982).
7. Z. AGUR and M. KERSZBERG, Amer.Nat., **129**, 6 (1987), 862-875.
8. R.ROSEN, Life Itself, A Comprehensive Inquiry Into the Nature, Origin, and Fabrication of Life (Columbia University Press, New York, 1991).
9. E.LASZLO, The Interconnected Universe: Conceptual Foundations of Transdisciplinary Theory (World Scientific Publishing Co., New Jersey, 1995).
10. A.LIMA-DE-FARIA, Evolution without Selection.Form and Function by Autoevolution (Elsevier, New York, 1988).
11. E.LASZLO, The Interconnected Universe: Conceptual Foundations of Transdisciplinary Theory (World Scientific Publishing Co., New Jersey, 1995).
12. E.LASZLO, Evolution: The Grand Synthesis (New Science Library, Boston, 1987).

13. I.PRIGOGINE, Order Out of Chaos: Man's New Dialogue With Nature (Bantam Books, New York:, 1984).
14. V.CSÁNI, Evolutionary Systems: A General Theory of Evolution (Duke University Press, Durham, NC.,1989).
15. The New Evolutionary Paradigm, (**2**) - keynote volume of the series The World Futures General Evolution Studiers, edited by E.Laszlo (Gordon and Breach Science Publishers, New York:, 1991).
16. J.GHARAJEDAGHI, Toward a Systems Theory of Organization. (Intersystems Pub., Seaside, CA, 1985).
17. M.TUGAN-BARANOVSKYI, The Annals of the Ukrainian Academy of Arts and Sciences in the United States, **XIII**, 35-36, (1973-1977), 189-208.
18. E.MAYR, The Growth of Biological Thought (The Belknap Press of Harvard University, Cambridge,1982).
19. A.KATSENELINBOIGEN, Systems Research, **8**, 4 (1991), 77-93.
20. A.KATSENELINBOIGEN, Some New Trends in Systems Theory (Intersystems Publications, Seaside, CA, 1984).
21. A.KATSENELINBOIGEN, in Proceedings of the International Conference Systems Inquiring: Theory, Philosophy, Methodology, **I**, (Los Angeles, 1985), 275-283.
22. A.KATSENELINBOIGEN, Selected Topics in Indeterministic Systems (Intersystems Publications, Seaside, CA,1989).
23. I.DAVYDOVSKII, Causality in Medicine (Medgis, Moscow, 1962)
24. J.GRAHAM, Cancer Selection. The New Theory of Evolution (Aculeus Press, Lexington, VA, 1992).
25. J.SHAPIRO, Genetica, **86,** 1-3 (1992), 9-11.
26. B.ALBERTS, D.BRAY, J.LEWIS, M.RAFF, K. ROBERTS, and J. WATSON, Molecular Biology of the Cell (Garland Publ., New York, 1989).
27. J.KOZA, Genetic Programming II (The MIT Press, Cambridge, MA, 1994).
28. J.STIGLITZ, in Handbook of Development Economics, edited by H. Chenery and T. Srinivasan. (Elsevier Science Publ., Amsterdam, 1988).
29. A.BERGSTROM, The Construction and Use of Economic Models (The English Universities Press, London, 1967).
30. J.SCHUMPETER, The Theory of Economic Development (Harvard University Press, Cambridge, MA, 1934).

31. K.POPPER, The Open Universe. An Argument for Indeterminism (Rowman and Littlefield, Totowa, 1982).
32. R.ATKIN, Int. J. Man-Machine Studies, **4**, (1972),139-167.
33. J.GLEICK, Chaos (Viking, New York, 1987).
34. A.KATSENELINBOIGEN, in Proceedings 5th IEEE International Symposium on Intelligent Control, (Philadelphia, 1990), 98-103.
35. R.HOFFFMAN, V. TORRENCE, Chemistry Imagined: Reflections on Science (Smithsonian Institution Press, Washington, 1993).
36. A.KATSENELINBOIGEN, Quarterly Journal of Ideology, **iii**, 1 (1980), 9-22.
37. E.MAYR, Populations, Species, and Evolution (Belknap Press of Harvard University Press, Cambridge, MA, 1970).
38. N.BERNSHTEIN, The Co-ordination and Regulation of Movements (Pergamon Press, New York,1967).
39. J.GHARAJEDAGHI in collaboration with R.ACKOFF, A Prologue to National Development Planning (Greenwood Press, New York, 1986).
40. E. MAYR, The Growth of Biological Thought (Belknap Press, Cambridge, MA, 1982).
41. E.PIELOU, An Introduction to Mathematical Ecology (Wiley-Interscience, New York, 1969).
42. R.ROSEN, Optimality Principles in Biology (Butterworths, London, 1967).
43. A.KATSENELINBOIGEN, Studies in Soviet Economic Planning (M.E. Sharpe Publ., White Plains, NY, 1978).
44. Study Week on the Economic Approach to Development Planning (Pontifical Academicae Scientiarum Scripta Varia, Vatican,1963).
45. E.JANTSCH, Design for Evolution (George Braziller, New York, 1975).
46. M.AKAM, Philos. Trans. R. Soc. Lond., B Bio. Sci., **349**, 1329 (1995), 313-319.
47. F.NIETZSCHE, Thus Spake Zarathustra (The Modern Library, New York, 1976).
48. L. O'NEILL, M. MURPHY, R. GALLAGHER, Science, 2**63**, 5144 (1994), 181-183.
49. The Flammarion Book of Astronomy, under the direction of G. Flammarion and A. Danjon in collaboration with a group of astronomers (Simon and Schuster, New York, 1964).

50. A.SEMENOV-TAYN-SHANSKY, Manuscripts of the Emperor's Academy of Sciences, **5**, 1910.
51. V.BUNAK, and V.ALEKSEEV, Antropologiia i Genogeografiia (Akademiia Nauk SSSR, Institut Etnografii, Moscow, 1974).
52. W.GREGORY, Proc. Natl .Acad. Sci. USA, **20**, 1(1934), 1-9.
53. W.GREGORY, A. J. P. A., **XX**, 2 (1935), 123-152.
54. A.BYSTROV, Man's Past, Present, and Future. (Medgiz, Moscow, 1957).
55. *Ibid.*
56. N.BUSHMAKIN, The Kazan Medical Journal, **13**, 1913.
57. F.NIETZSCHE, Thus Spake Zarathustra (The Modern Library, New York, 1976).
58. A.KATSENELINBOIGEN, Vertical and Horizontal Mechanisms as a System Phenomenon (Intersystems Publ., Seaside, CA, 1988).
59. E.MAYR, The Growth of Biological Thought (The Bellknap Press, Cambridge, MA, 1982).
60. T.SONNEBORN, Sci. Am., **183**, 5 (1950), 30-39.
61. T.SONNEBORN, and M. SCHNELLER, Dev. Genet. **1**, 1 1979, 21-46.
62. S.LØVTRUP, Epigenetics (John Wiley and Sons, New York, 1974).
63. P.SHEPPARD, Natural Selection and Heredity (Hutchinson Pub., London, 1958).
64. GOODWIN, How the Leopard Changed Its Spots. The Evolution of Complexity (Charles Scribner's Sons, New York, 1994).
65. K.UMANSKY, The Role of Viruses in Nature (Znanie, Moscow, 1981).
66. N.VORONTSOV, I.S.P.S, **3**, 2 (1989), 173-189.
67. I.DAVYDOVSKII, General Human Pathology (Mir Publishers, Moscow, 1971).
68. S.KAUFFMAN, The Origins of Order. Self Organization and Selection in Evolution (Oxford University Press, New York, 1993).
69. F.HAPGOOD, Why Males Exist: An Inquire Into the Evolution of Sex (William Morrow, New York, 1979).
70. C.WILLS, Discover, 13 (1992), 22-24.
71. A.GIBBONS, Science, **267**, 5206 (1995), 1907-1908.
72. N.ANGIER, N. Y. Times, January 25, 1994.
73. K.UMANSKY, The Role of Viruses in Nature (Znanie, Moscow, 1981).
74. J.MARX, Science, **261**,5127 (1993), 1385-1387.

75. J.WOLFF, R.MALONE, P.WILLIAMS, W.CHONG, G.ACSADI, A.ANI, P.FELGNER, Science, **247**, 4949 (1990), 1465-1468.
76. G.ACSADI, Nature, **52**, 1991, 815-818.
77. J.HARRIS, FEBS Letters, **95**, 1-3,1991, 3-4.
78. L.DUDKIN, I.RABINOVICH, I.VAKHUTINSKY, Iterative Aggregation Theory (Marcel Dekker, New York, 1987).
79. Principles of Self-organization, edited by H. Von Foerster and G. Zopf, Jr. (Pergamon Press, New York, 1962).
80. H.MATURANA and F.VARELA, Autopoiesis and Cognition: The Realization of the Living (D. Reidel Pub. Co., Boston, 1980).
81. Autopoiesis, a Theory of Living Organizations, edited by M.Zeleny (North Holland, New York, 1981).
82. M.TSEITLIN, Studies in Automata Theory and Simulation of Biological Systems (Nauka, Moscow, 1969).
83. V.VARSHAVSKY and D.POSPELOV, Orchestra Without a Conductor (Nauka, Moscow, 1984).
84. C.DUDDINGTON, Evolution in Plant Design (Faber, London,1969).
85. C.DUDDINGTON, Evolution and Design in the Plant Kingdom (Thomas Crowell Co., New York, 1969).
86. C.DARWIN, The Variation of Animals and Plants Under Domestication (Orange Judd, New York, 1868).
87. C.VILLEE, Biology (Saunders, Philadelphia, 1977).
88. J.GAIRNS, J.OVERBAUGH, S.MILLER, Nature, **335,** 6186 (1988), 142-145.
89. P. FOSTER, Annu. Rev. Microbio.,**47**, (1993), 467-504.
90. D.THALER, Science, **264**, 5156 (1994), 224-225.
91. R.HARRIS, S.LONGERICH, S.ROSENBERG, Science, **264**, 5156 (1994), 258-260.
92. N.AKOPYANTS, K.EATON, D.BERG, Infect. Immun., **63**, 1 (1995), 116-121.
93. E.CULOTTA, Science, **265**, 5170 (1994), 318-319.
94. J.RADICELLA, P.PARK, M.FOX, Science, **268**, 5209 (1995), 418-420.
95. T.GALITSKI, J.ROTH, Science, **268**, 5209 (1995), 421-423.
96. J.SHAPIRO, Science, **268**, 5209 (1995), 373-374.
97. G.SIMPSON, Tempo and Mode in Evolution (Columbia University Press, New York, 1944).
98. J. ENDLER, in Oxford Surveys in Evolutionary Biology, edited by R.Dawkins and M.Ridley, 3 (1986).

99. A. ROSENBERG, The Structure of Biological Science (Cambridge University Press, New York, 1985).
100. A.GEORGIEVSKY, Problemy Preadaptatsii (Nauka, Leningrad, 1974).
101. F.DOOLITTLE, and C.SAPIENZA, Nature, **284**, 5757 (1980), 601-603.
102. E.MAYR, Populations, Species, and Evolution (Belknap Press of Harvard University Press, Cambridge, MA, 1970).
103. E.MAYR, The Growth of Biological Thought (Belknap Press of Harvard University Press, Cambridge, 1982).
104. N.ELRIDGE, and S.GOULD, in Models of Biology, edited by T. Schopf and J. Thomas (Freeman, Cooper, San Francisco, 1972), 82-115.
105. J.HONDA, The World of Origami (Japan Publications Trading Co., Tokyo,1965).
106. Proceedings of the Twenty-fifth Annual North American Meeting of the Society for General Systems Research, (Toronto, Canada, 1981), 215-224.
107. R.LEWIN, Science, **224**, 4655 (1984), 1327-1329.
108. E.TARASOV, Physical Aspects of the Problems of Biological Evolution (Nauka, Moscow, 1979).
109. L.DODERLEINE, Biologisches Centralblatt, Bd. 7, 1888.
110. L.BERG, Nomogenesis, or Evolution Determined by Law (M.I.T. Press, Cambridge, MA, 1969).
111. H.OSBORN, Origin and Evolution of Life, 1916.
112. Mobile Genetic Elements, edited by J. Shapiro (Academic Press, New York , 1983).
113. J. SHAPIRO, Genetica, **84**, 3-4, 1991, 3-4.
114. E. DAVIDSON, Gene Activity in Early Development (Academic Press, Orlando, 1986).
115. J. SHAPIRO, BioEssays, **4**, 1 (1992), 791-792.
116. J. MARX, Science, **240**, 4854 (1988), 880-882.
117. B.LEWIN, Genes IV (Oxford University Press, New York, 1990).
118. R. NOWAK, Science, **263**, 5147 (1994), 608-609.
119. B. LEWIN, Gene Expression-volume 2: Eucariotic Chromosomes (Wiley, New York, 1980).
120. C. VILLEE, Biology (W.B.Saunders, Philadelphia, 1977).
121. R.DAWKINS, The Selfish Gene (Oxford University Press, New York, 1989).

122. F. DOOLITTLE, C. SAPIENZA, Nature, **284**, 5757 (1980), 601-603.
123. L. ORGEL, F. CRICK, Nature, **284,** 5757 (1980), 604-607.
124. F. EIJGENRAAM, Science, **256**, 5058 (1992), 730.
125. F.FLAM, Science, **266**, 5189 (1994), 320.
126. R.MANTEGNA, S.BULDYREV, A.GOLDBERGER, S.HAVLIN, C.PENG, M.SIMONS, H.STANLEY, Phys. Rev. Lett., **73**, 23 (1994), 3169-3172.
127. B.MCCLINTOCK, The Discovery and Characterization of Transposable Elements (Garland, New York, 1987).
128. J.SHAPIRO, Genetica, **86**,1-3 (1992), 109-111.
129. E.FOX KELLER, A Feeling for the Organism. The Life and Work of Barbara McClintock (W.H.Freeman and Co., New York, 1983).
130. N.FEDOROFF, in Mobil Genetic Elements (Academic Press, New York, 1983),1-57.
131. E.FOX KELLER, A Feeling for the Organism. The Life and Work of Barbara McClintock (W.H.Freeman and Co., New York, 1983).
132. New Perspectives on Evolution, edited by L.Warren and H.Korpowski (Wiley-Liss, New York, 1991).
133. M.KIDWELL, K. PETERSON, in New Perspectives on Evolution edited by L.Warren and H.Korpowski (Wiley-Liss, New York, 1991), 139-154.
134. J.SHAPIRO, BioEssays, **4**, 1 (1992), 791-792.
135. J.TRAVIS, Science, **257**, 5072 (1992), 884-885.
136. R.VON STERNBERG, G.NOVICK, G.GAO, R.HERRARA, Genetica, **86**, 1-3 (1992), 215-246.
137. L.ADLEMAN, Science, **266**, 5187 (1994), 1021-1024.
138. R.LIPTON, Science, **268,** 5210 (1995), 542-545.
139 G.KOLATA, N. Y. Times, April 11, 1995.
140. D.GIFFORD, Science, **266**, 5187 (1994), 993-994.
141. K.KELLY, N. Y. Times, May 15,1995.
142. L.BRUNDIN, Syst. Zool., **35**, 4, (1986), 602-607.
143. E.MAYR, Populations, Species, and Evolution (Belknap Press of Harvard University Press, Cambridge, MA, 1970).
144. S.GOULD, and E.VRBA, Paleobiology, **8**, 1 (1982), 4-15.
145. A.ROSENBERG, The Structure of Biological Science (Cambridge University Press, New York, 1985).
146. O.ABEL, Palaobiologie und Stammesgeschichte (Arno Press, New York, 1980).
147. J.SHAPIRO, Science, **269**, 5209 (1995), 286-287.

148. B.ALBERTS, D.BRAY, J.LEWIS, M.RAFF, K.ROBERTS, and J. WATSON, Molecular Biology of the Cell (Garland Publ., New York,1989).
149. E.LURIA, Patterns on the Glass (Znanie, Moscow,1982).
150. R.SERVICE, Science, **263**, 5152 (1994), 374.
151. K.WHITE, M.GRETHER, J.ABRAMS, L.YOUNG, K.FARRELL, H.STELLER, Science, **264**, 5159 (1994), 677-683.
152. M.SERRANO, G.HANNON, and D.BEACH, Nature, **366**, 6456 (1993), 704-707.
153. J.FOLKMAN, Y.SHING, Adv. Exp. Med. Biol., 313, (1992), 355-364.
154. D.DARLING, D.TARIN, New Scientist, **127**, 1726 (1990), 50-53.
155. V.ZAKIAN, Science, **270**, 5242 (1995),1601-1607.
156. N. ANGIER, N. Y. Times, June 9, 1992.
157. H.MÜLLER, The Collecting Net, **13**, 8 (1938), 181-195, 198.
158. B.MCCLINTOK, Proc. Natl .Acad. Sci. USA, **25**, 405 (1939).
159. Advanced Cell Biology, edited by L.Schwartz and M.Azar (van Nostarnd Reinhold Co., New York, 1981).
160. A. OLOVNIKOV, J.Theoret. Biol., **41**, 1 (1973),181-190.
161. C.HARLEY, H.VAZIRI, C.COUNTER, and R.ALLSOPP, Exp. Gerontol., **27**, 4 (1992), 375-382.
162. M.COUNTER, A.AVILION, C.LEFEUVRE, N.STEWART, C.GREIDER, C.HARLEY, and S.BACCHETTI, EMBO J., **11**, 5 (1992), 1921-1929.
163. K.WHITE, M.GRETHER, J.ABRAMS, L.YOUNG, K.FARRELL, H.STELLER, Science, **264**, 5159 (1994), 677-683.
164. N. ANGIER, N.Y.Times, May 17, 1994.
165. V.DILMAN, Development, Aging, and Disease: A New Rationale for an Intervention Strategy (Langhorne, Pa., Harwood Academic Publ., 1994).
166. J. SHAPIRO, Genetica, **84**, 3-4 (1991), 3-4.
167. The Socialist Price Mechanism, edited by A. Abouchar (Duke University Press, Durham, NC., 1977),171-183.
168. C.HARLEY, H.VAZIRI, C.COUNTER, and R.ALLSOPP, Exp. Gerontol., **27**, 4 (1992), 375-382.
169. J.EGELHOFF, J.SPUDICH, Trends Genet., **7**, 5, (1991), 161-166.
170. A. KONDRASHOV, J. Hered., **84**, 5 (1993), 372-387.
171. J.SMITH, The Evolution of Sex (Cambridge University Press, New York, 1978).

172. A.FAUSTO-STERLING, The Sciences, **33**, March/April 1993, 20-25.
173. D.CREWS, Sci.Am., **270**, 1 (1994), 108-114.
174. D.ACKERMAN, N. Y. Times, December 17, 1995.
175. E.ADELBERG, in The Origin and Evolution of Sex, edited by H.Halvorson and A.Monroy (Alan R.Liss, Nerw York, 1985), 87-88.
176. E. MAYR, The Growth of Biological Thought. Diversity, Evolution, and Inheritance (Belknap Press of Harvard University Press, Cambridge, MA, 1982).
177. E. WILSON, Sociobiology. The New Synthesis. (Belknap Press of Harvard University Press, Cambridge, MA, 1975).
178. A.BURT, D.CARTER, G.KOENIC, T.WHITE, J.TAYLOR, Proc.Natl.Acad.Sci.USA, **93**, 2 (1996). 770-773.
179. T. SONNEBORN, Sci. Am., **83**, 5, (1950), 30-39.
180. H.BERNSTEIN, H.BYERLY, F. HOPF, and R.MICHOD in The Origin and Evolution of Sex, edited by H.Halvorson and A.Monroy (Alan R.Liss, New York, 1985), 29-45.
181. E. WILSON, Sociobiology. The New Synthesis (Belknap Press of Harvard University Press, Cambridge, MA, 1975).
182. E. ECKHOLM, N. Y. Times, March 25, 1986.
183. M. ROSE, F. DOOLITTLE, New Scientist, **98**, 1362 (1983), 787-789.
184. A. ANDERSON, Science, **257**, 5068 (1992), 324-326.
185. L. HURST, Nature, **354**, 6348 (1991), 23-24.
186. G.HURST, L.HURST, M.MAJERUS, Nature, **356**, 6371 (1992), 659-660.
187. L.HURST, W.HAMILTON, Proc. R. Soc. Lond., B Biol.Sci., **247**, 1320 (1992), 189-194.
188. L.HURST, Proc. R. Soc. Lond., B Biol.Sci., **248**, 1322 (1992), 135-140.
189. L.HURST, R.HOEKSTRA, Nature, **367**, 6463 (1994), 554-557.
190. V.GEODAKIAN, Nauka i Zhizn, 3 (1966), 99-105.
191. S.SCHRAG, A.MOOERS, G.NDIFON, and A.READ, Am. Nat., **143**, 4 (1994), 636-655.
192. R.HOEKSTRA, in The Evolution of Sex and its Consequences, (Basel, Birkhauser, 1987), pp. 59-91.
193. R.HOWARD, C.LIVELY, Nature, **367**, 6463 (1994), 554-557.
194. C.VILLEE, Biology (Saunders, Philadelphia, 1977).

195. A.FORSYTH, A Natural History of Sex (Charles Scribner's Sons, New York, 1986).
196. A.SURANI, Nat. Genet.,**11**, 2 (1995), 111-113.
197. L.STRAIN, J.WARNER, T.JOHNSTON, D.BONTHRON, Nat. Genet, **11**, 2 (1995), 164-169.
198. J. HALDANE, Annals of Eugenics, **13**, (1946-1947), 262-271.
199. T. MIYATA, H. HAYASHIDA, K. KUMA, K. MITSUYASU, T.YASUNAGA, Cold Spring Harb. Symp. Quantit. Biol, **52**, (1987), 863-867.
200. A.KONDRASHOV, Nature, **369**, 6476 (1994), 99-100.
201. R.REDFIELD, Nature, **369**, 6476 (1994), 145-146.
202. S.SPENCE, D.GILBERT, D.SWING, N.COPELAND, and N.JENKINS, Mol. Cell. Biol., **9**, 1 (1989), 177-184.
203. B.BACCETTI, A.BENEDETTO, A.BURRINI, G.COLLODEL, E.CECCARINI, N.CRISA, A.DICARO, M.ESTENOZ, A.GARBUGLIAR, A.MASSACE, P.SIPIOMBONI, T.RENIERI, D.SOLAZZO, J. Cell Biol., **127**, 4 (1994), 903-914.
204. C. VILLEE, Biology (Saunders, Philadelphia,1977).
205. L.SHIMMIN, B.CHANG, W.LI, Nature, **362**, 6422 (1993), 745-747.
206. N. ANGIER, N. Y. Times, May 17, 1994.
207. YU.VASILIEV, I.GELFAND, Interaction between Normal and Neoplastic Cells and the Environment (Nauka, Moskow, 1981).
208. V.DILMAN, The Grand Biological clock (Mir Publ., Moscow, 1989).
209. J.LADIK, W. FÖRNER, The Beginnings of Cancer in the Cell (Springer-Verlag, New York, 1994).
210. J.GRAHAM, Cancer Selection (Aculeus Press, Lexington, VA, 1992).
211. F.LEDILY, J.BILLARD, C.HUAULT, C.KEVERS, T.GASPAR, In Vitro Cell.&Dev. Biol. Plant, **29**, 4 (1993),149-154.
212. J. LADIK, W. FÖRNER, The Beginnings of Cancer in the Cell (Springer-Verlag, New York, 1994).
213. V.VDOVICHENKO, A.PODOROZHNYI, V.MAKARA, Klinicheskaja Medicina, **69**, 8 (1991), 102.
214. A.CAMERIERI, E.COSTA, F.SCARANO, R.COLACINO, G. Ital. Cardiol., **2**, 9 (1992), 1093-1097.
215. Y.MORIYAMA, H.SAIGENZI, S.SHIMOKAWA, Y.UMEBAYASHI, H. TOYOHIRA, M.HASHIGUCHI, A.TAIRA, J JPN Assoc. Thorac. Surg., **41**, 3 (1993), 367-371.

216. A.TURNER, and N.BATRICK, Int. J. Cardiol., **40**, 2 (1993), 115-119.
217. S.KOSLING, H.SCHULZ, J.STEINDORF, H.WEIDENBACH, Rofo Fortschr Geb Rontgenstr Neuen Bildgeb Verfahr, vol. **158**, 2 (1993), 180-182.
218. J.WYKE, Br. J. Cancer, **62**, 2 (1990), 341-347.
219. E. COWDRY, Cancer Cells (W.B.Saunders, Philadelphia,1955).
220. A.KATSENELINBOIGEN, Selected Topics in Indeterministic Systems, (Intersystems Publications, Seaside, CA, 1989).
221. E.WALKER, Acta Biotheor., **40**, 1 (1992), 31-40.
222. E.COWDRY, Cancer Cells (W.B.Saunders, Philadelphia,1955).
223. The Cancer Dictionary. (Facts on File, New York, 1992).
224. Britannica, **5**, 15 edition.
225. B.ALBERTS, D.BRAY , J.LEWIS, M.RAFF, K.ROBERTS, and J.WATSON, Molecular Biology of the Cell (Garland Publ., New York,1989).
226. S.BLAKESLEE, N. Y. Times, 17 May, 1994.
227. J.GRAHAM, J. Theoret. Biol., **101**, 4 (1983), 657-659.
228. R.SHIMKE, S.SHERWOOD, A.HILL, R.JOHNSTON, Proc. Natl.Acad. Sci. USA , **83**, 7 (1986), 2157-2161.
229. T.KRONTIRIS, B.DEVLIN, D.KARP, N.ROBERT, N.RISCH,N. Engl. J. Med., **329**, 8 (1993), 517-523.
230. G. GEORGIEV, Nauka i Zhizn, 5 (1981), 92-96.
231. J.SHAPIRO, Genetica, **6**, 3 (1992), 109-111.
232. G.ZAJICEK, Med. Hypotheses, **6**, 6 (1980), 665-670.
233. J.GHARAJEDAGHI, in collaboration with R.ACKOFF, A Prologue to National Development Planning (Greenwood Press, New York, 1986).
234. J.MARX, Science, **259**, 5096 (1993), 760-761.
235. R.WILLIS, The Spread of Tumors in the Human Body (Butterworth&Co, London, 1952).
236. C.COUNTER, H.HIRTE, S.BACCHETTI, and C.HARLEY, Proc. Natl. Acad. Sci. USA, **91**, 8 (1994), 2900-2904.
237. M.NASH, Time, **143**, 17 (1994), 54-61.
238. T.IMAIZUMI, Cancer and Field (Tokyo, 1982).
239. E.SPEIR, R.MODALI, E.HUANG, M.LEON, F.SHAWL, T.FINKEL, S.EPSTEIN, Science, **265**, 5170 (1994), 391-394.
240. G.KOLATA, N. Y. Times, May 11, 1993.

241. L.MECKLER, Z. Vses. him. obschestva im. D.I.Mendeleeva, **25**, 3 (1980), 333-357.
242. J.MARX, Science, **261**, 5127 (1993), 1385-1387.
243. M.BAYER, H.KAISER, M.MICOZZI, In Vivo, **8,** 1 (1994), 3-15.
244. Cancer, Principles & Practice of Oncology. (J.B. Lippincot, Philadelphia, 1993).
245. L.MECKLER, Z. Vses. him. obschestva im. D.I.Mendeleeva, **13** (1968).
246. L.MECKLER, Z. Vses. him. obschestva im. D.I.Mendeleeva, **25**, 3 (1980), 333-356.
247. J.MARX, Science, **246**, 4928 (1989), 326-328.
248. V.DILMAN, The Grand Biological clock (Mir Publ., Moscow, 1989).
249. Pathological Anatomy of Dirsorders in Fetus and Chidlren, edited by T. Ivanovskaya and L. Leonova (Meditsina, Moscow, 1989).
250. S.ANTOHI, Virologie, **33**, 3 (1982), 241-249.
251. II.TEMIN, Cancer Res., **48**, 7 (1988), 1697-1701.
252. NOWELL, P., Proc. Am. Phil. Soc., 139, 1 (1995), 32-43.
253. J.WYKE, Br. J. Cancer, **62**, 2 (1990), 341-347.
254. C.ECKERT, Med. Hypotheses, **9**, 1 (1982), 87-94.
255. K.SETALA, Med. Hypotheses, **15**, 3 (1984), 209-230.
256. K.GROSSGEBAUER, Biosystems, **16**, 3-4 (1983-1984), 253-258.
257. O.WARBURG, The Metabolism of Tumours (R.R.Smith, New York, 1931).
258. Z.CHEREISKY, Khimiia i Zhizn, 3 (1973), 80-82.
259. G.COLATA, N. Y. Times, December 22,1992.
260. R.SERVICE, Science, **263**, 5152 (1994),1374.
261. J.MARX, Science, **261**, 5127 (1993), 1385-1387.
262. J.MARX, Science, **264**, 5157 (1994), 344-345.
263. J.MARX, Science, **266**, 5189 (1994), 1321-1322.
264. Cancer, Principles& Practice of Oncology (J.B. Lippincot, Philadelphia,1993).
265. M.HALPERN, S.McMAHON, J. Immunol., **138**, 9 (1987), 3014-3018.
266. M.HALPERN, D.EWERT, J.ENGLAND, Virology, **175,** 1 (1990), 328-331.
267. M.STRACKE, S.AZNAVOORIAN, M.BECKNER, L.LIOTTA, E. SCHIFFMANN, EXS, 59 (1991), 147-162.

268. D.DARLING and D.TARIN, New Scientist, **127**, 1726 (1990), 50-53.
269. A.KRIEG, A.YI, S.MATSON, T.WALDSCHMIDT, G.BISHOP, R.TEASDALE, G.KORETZKY, D.KLINMAN, Nature, **374**, 6522 (1995), 546-549.
270. V.DILMAN, The Grand Biological Clock (Mir Publ., Moscow, 1989).
271. Tokyo Metropolitan Geriatric Hospital, JPN J. Geriatr., **30**, 1 (1993), 35-40.
272. Cancer Facts & Figures - 1991 (American Cancer Society).
273. D.ILSON, G.BOSL, R.MOTZER, E.DMITROVSKY, Hematol. Oncol. Clin. North Am., **5**, 6 (1991), 1271-1283.
274. S.SPENCE, D.GILBERT, D.SWING, N.COPELAND, and N.JENKINS, Mol. Cell. Biol., **9**, 1 (1989), 177-184.
275. B.BACCETTI, A.BENEDETTO, A.BURRINI, G.COLLODEL, E.CECCARINI, N.CRISA, A.DICARO, M.ESTENOZ, A.GARBUGLIAR, A.MASSACE, P.SIPIOMBONI, T.RENIERI, D.SOLAZZO, J. Cell Biol., **127**, 4 (1994), 903-914.
276. E. MAYR, Populations, Species, and Evolution (Belknap Press of Harvard University Press, Cambridge, 1970).
277. L.MECKLER, Z. Vses. him. obschestva im. D.I.Mendeleeva, **25**, 3 (1980), 333-356.
278. W.LANE, L.COMAC, Sharks don't Get Cancer (Avery Publ., Garden City Park, NY, 1993).
279. T.BEARDSLEY, Sci. Am., **269**, 4 (1993)
280. C.CILLO, Invasion Metastasis, **14**, 1 (1994-1995), 38-49.
281. N.VORONTSOV, Everyman's Science, April-May (1989), 53-54.
282. O.OLIVEROS, E.YUNIS, Cancer Genet. Cytogenet., **82**, 2 (1995), 155-160.
283. W.DILLS, Parasitology, **107**, Supplement (1993), 177-186.
284. *ibid.*
285. T.DELANGE, Proc. Natl .Acad. Sci. USA, **91**, 8 (1994), 2882-2885.
286. E.ROSENTAL, N. Y. Times, December 5, 1989.
287. J.MARX, Science, **248**, 4953 (1990), 442-443.
288. R. von WEITZAKER, Strana i Mir, **3**, 51 (1989), 151-161.
289. J.YARNOLD, Eur. J. Cancer **28**, 1 (1992), 251-255.
290. M.BAKHTIN, Problems of Dostoyevsky's Poetics (University of Minnesota Press, Minneapolis, 1984).
291. G.TAUBES, Science, **260**, 5110 (1993), 884-886.

292. J.MARKOFF, <u>N. Y. Times</u>, January 8, 1995.
293. S.LØVTRUP, <u>Rivista di Biologia,</u> **75**, 2 (1982), 231-272.
294. E.MIKULINA, <u>Socialist Competition of the Masses</u> (Cooperative Pub., Moscow,1932).
295. J.STALIN, in <u>Collected Works</u>, **12**, (Gospolitizdat, Moscow, 1951), 112-115.

# INDEX

Ackoff, R., 52
Adaptation, 13, 41, 65, 66, 67, 77, 79, 84, 85, 87, 105, 109, 146, 156, 163
   preadaptation, 84–86
   proadaption, 84–86
Adleman, L., 3, 102–104
Aflatoxin, 181
Age, 112, 114, 115, 175, 176, 182, 184, 185, 190
AIDS, 171, 174, 185
Angiogenesis, 180
Apoptosis, 45, 110, 116, 170, 174
Arrow, K., 158
Autonomous, 2, 3, 32, 33, 35, 37, 61, 79, 94, 97, 108, 169, 173

Beach, D., 110
Beardsley, T., 181, 182
Berg, L., 92
Bergson, H., 3
Bishop, M., 161, 165
Black box, 3, 54, 91
Blakeslee, S., 153
Bosch, H., 37

Cancer:
   definition, 5, 6, 8, 11, 12, 45, 47, 150–154, 164
   malignancy, 5, 147, 148, 151, 152, 156, 159, 160, 164, 165, 167–169, 171, 174, 175, 182–184, 191
   metastasis, 5, 114, 116, 144, 148, 151, 161, 169, 171, 173, 174, 178, 183–185, 189, 191
   sarcoma, 47, 160, 185
   universality, 146–148
Cells:
   differention, 108, 109, 168, 169, 173, 175, 183
   germs, 2, 4, 5, 8–10, 12, 36, 44, 62, 68, 71, 73, 76–84, 86, 92–94, 107, 109, 110, 115, 118–126, 128, 130, 133, 135–138, 153, 162, 163, 165, 168, 176–178, 182, 183, 187, 188, 190
   somatic, 2, 4, 8, 10, 12, 36, 62, 71, 76–81, 83, 84, 94, 107, 113, 116, 118, 136–138, 145, 153–155, 162–165, 169, 173, 177, 182, 183, 187, 190
   stems, 108
   zygote, 118, 128, 129, 153
Chemicals:
   glucose, 169, 184
   lactic acid, 169
   oxygen, 33, 109, 169, 173
Chereisky, Z., 169
Comac, L., 181
Copernicus, 71
Csáni, V., 7
C-value paradox, 25, 95, 115, 188

Darwin, C., 11, 21, 68, 78
Davydovskii, I., 16
Dawkins, R., 97
Delbruck, M., 85
Deviants, 12, 41–44, 46, 47, 114, 158, 160, 172, 174
   bandits, 42, 44, 45, 47
   innovators, 3, 5, 26, 27, 33, 36, 37, 41, 42, 44–46, 108, 109, 111, 113, 114, 152, 156, 158–160, 164, 171–174
   radicals, 5, 42–44, 46, 47, 151, 152, 158–160, 164, 172
   reformers, 42, 46

Dilman, V., 115, 144, 165, 176
Directivness, 3, 53, 170
   extrapolation, 31, 53, 59
   law, 59, 70, 71, 91, 92, 105
   teleological, 53
   telological, 53, 56
   telenomic, 53, 91
   teleomatic, 53
DNA-molecule "computer", 4, 102–105
Doderleine, L., 92, 105
Dynamics:
   survival, 7, 11, 12, 17, 21, 24, 25, 28, 40, 41, 46, 52, 53, 56–59, 62, 66, 71, 75–77, 85, 90–92, 97, 106, 112, 120, 154, 155, \ 157, 159, 160, 194
   viability, 7, 52, 75
   growth, 7, 52, 53, 57, 62, 75, 76, 90, 112, 120, 164, 165, 175
   development, 2–4, 6–10, 13, 21, 22, 24–26, 28–32, 34, 35, 37, 38, 40, 41, 44–47, 51–57, 61–63, 67, 69, 72, 73, 75–77, 79, 82, 86–93, 97, 105, 106, 108–110, 112, 114–116, 119, 122, 123, 128, 132, 133, 135, 137, 138, 153–155, 158, 159, 162–170, 172–175, 178, 180, 188–190, 192
Embryonic, 52, 56, 110, 112, 114, 131, 135, 163,164, 167, 169, 175, 190
Equilibrium, 23, 24, 54, 55, 87, 11
Euclid, 37
Evolution:
   creationism, 54, 56
   general theory, 7, 8
   neo-Darwinism, 11, 93, 191, 192
   punctuated, 87
   vitalism, 3

Fausto-Sterling, A., 122
Filonenko, G., 47
Flammarion, C., 58
Fox Keller, E., 101
Frampton, C., 181

Galois, E., 158
Garfield, E., 193
Generative system, 62, 64, 188
   aristogenesis, 92
   chromosomes, 3, 5, 56, 63–68, 73, 86, 98, 100, 109, 111–113, 126, 128, 135, 164, 165, 168, 184
   cytoplasm, 63, 64, 72, 73, 113, 128, 129
   DNA, 4, 62, 67, 72, 73, 85, 93–97, 99, 100, 102–106, 109, 127, 136, 140, 147, 151, 153, 156, 165, 168, 170–172, 174, 181, 191
   genome, 2–4, 9, 11, 25, 29, 36, 44, 45, 63, 65, 72, 73, 79–81, 83, 84, 86, 88, 91, 93–97, 99, 101–104, 106, 108, 113, 116, 126, 147, 156, 162, 164, 165, 167, 168, 170, 173, 188
   hormones, 63, 72,123
   HOX genes, 56, 167, 183, 184
   mitochondria, 63, 128, 129, 168
   "non-Mendelian inheritance", 64
   $p\,16$ gene, 170
   $p\,53$ gene, 161, 170
   proteins, 56, 73, 95, 96, 100, 103, 106, 164, 165, 168, 171, 180
   RNA, 62, 67, 72, 100, 103, 112, 140, 168
   selfish DNA, 4, 25, 88, 95–97, 99–102, 104, 105, 136, 155
Genetic programming, 15–17
Geodakian, V., 130
Gharajedaghi, J., 10, 52
Gödel, K., 157
Gogh, V., 37
Goodwin, B.,
Gould, S., 87, 105
Graham, J., 16
Gregory, W., 59
Gross, N., 42

Hamilton,W., 128,
Harshbarger, J., 181
Hermaphrodite, 119, 120, 122–124, 126, 130, 132, 138
   full, 122, 123, 138
   herms, 122, 123,
   merms, 123
   ferms, 123
   chimera, 124
Herrera, R., 102
Hickey, D., 127
Hurst, L., 128, 129
Hypothesis, 161

Indeterminism, 13, 29, 30, 39, 40, 70
  beauty, 31
  predisposition, 4, 29–32, 35, 90, 105, 152, 153, 157, 165, 168, 176, 181–183, 190
  tunnel process, 4, 7, 13, 28–30, 35, 36, 69, 86–88, 91, 94, 105, 115, 127, 128, 133, 136, 163, 188
Immune system, 5, 12, 44–47, 104, 111, 115, 145, 151, 159, 165, 168, 171, 172, 176, 180–183, 185, 190
Isomorphism, 13, 34

Kandinsky, W., 37
Kauffman, S., 68, 73
Keynes, J.M., 159
Koprowski, H., 172

Lane, W., 181
Laszlo, E., 7
Living beings:
  ass, 73
  drosophila, 66, 101, 111, 129
  fish, 123, 179
  fungus, 125
  green algae *Chlamydomonas*, 128
  hinny, 73
  larva, 124
  mammals, 60, 72, 73, 96, 108, 126, 133, 135, 156, 180
  mare, 73
  mule, 73
  paramecium, 108, 125
  sharks, 180, 181
  snails, 122, 130,
  stallion, 73
  symbion pandora, 122, 124
Løvtrup, S., 164, 194

Machines, 11, 61, 81, 82, 103, 104, 104, 106, 158, 170, 194
Malthus, T., 12, 21
Marx, K., 159
Mayr, E., 49, 53, 64, 86, 87, 91
McClintock, B., 1, 93, 94, 100, 101, 111
McDonald, J., 102
Mechanisms of change:
  germatic, 76–79, 82, 83, 116, 118, 119, 137, 153, 188, 189
  hierarchies, 56, 64, 65, 67–69, 71, 77, 80–82, 90, 105, 108, 115, 129, 149
  horizontal, 9, 24, 26–28, 46, 54, 62, 64, 108, 167
  innovations, 3, 9, 21–27, 31–36, 39, 45, 46, 52, 63, 64, 69, 80, 86, 109, 116, 127, 133, 137, 162, 170, 172, 173, 175, 193
  internal, 2–6, 8, 9, 12, 13, 23, 28, 29, 33, 36, 51, 64, 65, 68, 74, 78, 84, 86–96, 98–100, 102–105, 109, 112, 115, 116, 135, 145, 150, 155, 162, 167, 168, 181, 182, 187, 188, 190
  mutations, 1, 2, 12, 16, 23, 24, 36, 66–71,73, 84, 85, 89, 91, 94, 100, 104–106, 109, 110, 120, 128, 134, 136, 140, 150, 151, 155, 156, 165, 170, 184–186, 191, 194
  somatic, 2, 4–6, 11– 13, 47, 76–83, 107, 108, 111–114, 116, 118, 137, 138, 146, 151, 153–155, 162–164, 167, 171, 172, 176, 182, 187– 191
  vertical, 9, 26, 27, 54, 62– 64
Meckler, L., 163, 165, 179
Microcosm, 44, 79
Moishezon, B., 37, 140, 194
Montesquieu, C., 38
Music, 13, 69, 70

National Cancer Institute, 172, 181
Natural philosophy, 6, 13, 14
Nietzsche, F., 57

Offspring, 9, 76, 118–122, 129, 139, 178
Organs and structures:
  bronchi, 179
  colon, 72, 183
  blood/lymph stream, 171
  cartilage, 180, 181
  duodenum, 179
  esophagus, 179
  heart, 60, 81, 166
  liver, 179
  lung, 179, 183, 184
  stomach, 179, 184
  trachea, 179, 184
Origami, 13, 88, 89, 106
Osborn, H., 92
"Overman", 58

Parkinson's disease, 141, 143
Pathology, 1, 2, 5, 6, 8, 11, 12,16, 41,
  42, 44, 45, 47, 66, 83, 93, 106, 107,
  113, 114, 116, 141, 143–146, 151,
  153, 155, 157, 161–164, 167–170,
  172, 173, 175, 182–184, 187–191,
  193
Pavlov, I., 51
Phenomenology, 3, 54, 59, 60, 91, 94,
  179
Plants, 5, 56, 57, 66, 76,77, 92, 96, 100,
  146–148, 164
Political system:
  judicial branch, 39, 132, 138, 189
  pluralistic mechanism, 14, 26, 38,
    41, 43, 46, 150
  separation of powers, 38, 39, 46, 47,
    55, 132, 138, 189
Programs, 4, 34–36, 65, 82, 83, 90, 94,
  115, 153, 159, 167, 168, 170, 175,
  188
  zero-level, 34, 35, 175, 182, 188,
    190
  first level, 34, 35, 90, 94, 95, 115,
    116, 175, 182, 188, 190
  second level, 34, 35, 90, 94, 95, 115,
    116, 175, 188
  third level, 34, 94, 95
Ptolemy, 71
Pythagoras 37

Radiation, 2, 36, 65, 84, 106, 108, 115,
  126, 134, 140, 152, 168, 173, 183,
  188
Redfield, R., 136
Regulatory mechanism, 4, 37, 73, 93,
  96, 99, 106, 108, 117, 155, 156,
  164, 169, 174, 184
Repair mechanism, 4, 49, 75, 93, 96, 99,
  106, 109, 110, 116, 126, 149, 156,
  162, 170, 174
Reproductive system:
  egg (ovum), 64, 112, 123, 126, 131,
    134, 135, 138, 139, 140, 178
  fragmentation, 76, 118, 120
  ovary, 72, 112, 123, 133, 134, 137,
    138, 177–179, 184, 189, 191
  parthenogenesis, 119, 122, 130, 135–
    137
  sperm, 111, 112, 123, 124, 130, 133–
    136, 139, 140, 151, 178
  testicles, 133, 134, 137, 138, 189
Research and Development, 22, 32–36,
  106, 133, 145, 187
Roosevelt, F., 159
Rose, M., 127
Rosen, R., 4
Rosenberg, S., 172

Schumpeter, J., 22, 24, 27
Schuster, G., 193
Self-organization, 73, 74
Sexes: 4, 6, 17, 39, 76, 77, 118–120,
    122, 124, 126–128, 131, 133, 135,
    136, 138–140, 176, 177, 188
  "compatibles", 5, 119, 120, 127, 129
  crossover, 5, 16, 17, 65, 73, 84, 118–
    122, 124, 125, 127, 128, 131,
    132, 137–139, 168, 189
  definition, 4, 5, 12, 118, 120–123,
    125, 131, 137, 188
  females, 39, 72, 123, 126, 130, 133,
    134, 137, 138, 183, 189, 190
  males, 39, 72, 123, 129, 130, 133–
    136, 138, 139, 183, 189, 190, 191
  multisexual, 5, 12, 39, 47, 76, 124,
  sex-convertible, 123, 124
Shapiro, J., 16, 85, 93, 101, 105, 116,
    156.
Sheppard, P., 64
Slonimsky, P., 98
Sonneborn, T., 64, 126
Spore, 8, 76, 77, 119, 120
Sternberg, von, R., 102
Systems approach, 6, 10, 12, 30, 149,
    155

Telomerase, 5, 112, 113, 116, 117, 149,
    159, 170, 173, 184, 185, 190
Telomers, 5, 67, 109–113, 149, 157,
    160, 170, 173, 185, 187
Transposons, 4, 5, 65, 94, 100–102, 104,
    106, 168
Tugan-Baranovsky, M., 12
Tumors, 16, 112, 147–149, 151, 157–
    159, 162, 164, 165, 169, 170, 171,
    173, 174, 179–181, 184

Umansky, K., 65,

Varmus, H., 161, 165
Viruses, 5, 63, 65–67, 72, 106, 108, 109, 113, 127, 134, 136, 140, 150, 152, 155, 164, 165, 168, 169, 171, 173, 180
Vogelstein, B., 162
Vorontsov, N., 66, 184

Williston, S., 59
Wilson, E., 125
Wistar Institute, 172